ENCYCLOPAEDIA
OF
MINERALOGY

ENCYCLOPAEDIA OF MINERALOGY

Volume 3

Bharat Singh

ANMOL PUBLICATIONS PVT. LTD.
NEW DELHI-110 002 (INDIA)

ANMOL PUBLICATIONS PVT. LTD.
Regd. Office: 4360/4, Ansari Road, Daryaganj,
New Delhi-110002 (India)
Tel.: 23278000, 23261597, 23286875, 23255577
Fax: 91-11-23280289
Email: anmolpub@gmail.com
Visit us at: www.anmolpublications.com

Branch Office: No. 1015, Ist Main Road, BSK IIIrd Stage
IIIrd Phase, IIIrd Block, Bengaluru-560 085 (India)
Tel.: 080-41723429 • Fax: 080-26723604
Email: anmolpublicationsbangalore@gmail.com

Encyclopaedia of Mineralogy

ISBN 978-81-261-3475-5 (Set)

PRINTED IN INDIA

Printed at AnVi Composers, New Delhi.

Contents

Preface

'Mineralogy' is a branch of science, under which we study about the minerals, which are obtained from the earth. All the minerals are found in the earth. We dig out the earth for getting minerals in the form of ores, which are impure. By the process of Metallurgy, we get pure elements. A few metals such as Au (gold), Ag (silver) and Pt (platinum) are found in pure form. A mineral is defined to be a naturally occurring, homogeneous crystalline solid. First, a mineral should be a solid; this requirement excludes all liquids and gases. A crystalline material is not merely a solid, but also possesses an orderly and periodic molecular structure. In recent years, however, a multitude of synthetic materials, which would otherwise satisfy the definition of a mineral, have been produced in the laboratory. These substances are rapidly gaining in importance in both industry and experiment, and it now seems logical to extend the definition of 'mineral'. Minerals are inorganic solids, which have a regular three-dimensional ordered structure of atoms.

Minerals are classified on the basis of their chemical composition and their isomorphism or similarity of properties, such as structural properties, physical properties and chemical properties.

In structural properties, we study about the structure of the crystal. Crystallographic property is one of the most characteristic properties of the crystals, and in this property we study about the shape of the crystals. Other structural property is 'optical properly', in which we study about the crystal structure, the kind of atoms present in the structure and their electronic arrangement. In physical properties, we study about the properties such as colour, taste, melting and boiling points, density, etc. In chemical properties, we study about the reactions of the elements/minerals with others and products after reactions and their relation with the reactants.

This eminent work, *Encyclopaedia of Mineralogy* has specific and realistic features about minerals in an interesting manner, which will be able to make a creative impression in the study of mineralogy. Expectedly, this work would be welcomed and liked by all.

— **Editor**

1

The Oxides

As is known, oxygen plays an outstanding role in the chemical processes in the earth's crust; this concerns both inorganic and organic mineralogy. Oxygen compounds, from simple oxides to various oxysalts, are sharply predominant throughout the earth's crust. We might recall that the mean oxygen content in the earth's crust is 47.0 per cent by weight.

The total weight of free oxides in the lithosphere (hydrosphere and atmosphere excepted) is about 17 per cent. The share of silica alone is 12.6 per cent. Iron oxides and hydroxides constitute 3.9 per cent. Other important oxides and hydroxides are those of aluminium, manganese, titanium, and chromium.

The atmosphere contains oxides in the form of carbon dioxide and water vapour which extend in appreciable quantities up to a height of 12 km from the surface of the earth. The hydrosphere, as its name implies, is chiefly water.

Genesis and Chemical Features of Compounds: The bulk of oxides and hydroxides of most diverse composition is concentrated in the uppermost parts of the earth's crust, i.e., on the border with atmosphere which contains free oxygen. The depth of intense penetration of free oxygen into the earth's crust is controlled mainly by the ground-water table. The crust of weathering, together with the oxidised zones of ore deposits, is the principal arena of chemical reactions which produce new formations, among which the oxides and hydroxides of metals play the dominant role.

An essential role is played in these processes not only by free oxygen of the air which penetrates from the surface into the crust of the earth, but also by percolating rain water which carries dissolved oxygen and carbon dioxide. Air-saturated rain water contains on the average from 25 to 30 cu cm of gas per litre. The gas consists of 30 per cent oxygen, 60 per cent nitrogen, and 10 per cent carbon dioxide. Thus rain water is heavily saturated with oxygen and particularly with carbon dioxide.

As the water percolates towards the ground-water table, its oxidising power gradually diminishes, the free oxygen being consumed in oxidation. This depletion is especially intense in the oxidation of sulphides and similar compounds, resulting at the first stage in formation of sulphates, arsenates, etc.

Also rather easily oxidised are oxygen compounds entering into the composition of minerals and ores containing low-valent metals, such as Fe^{2+}, Mn^{2+}, V^{3+}. In the course of oxidation these metals are replaced by ions of higher valency, but of smaller size, and hence the bonds of the crystal structure become weaker. In the end the oxidised crystalloids completely break down which is followed by the formation of new compounds, some soluble and some insoluble in water.

The salts, (sulphates, carbonates, etc.) forming during the first stage of these processes react with water, being subjected to *hydrolysis*, as a result of which a number of metal cations precipitate in the form of hydroxides of low water solubility.

It is well known that the behaviour of ions in solution depends on the ionic potentials, whose values are expressed in terms of the charge-to-ionic-radius ratio *(W : RI)*. The cations yielding poorly water-soluble hydroxides are, in the order of increasing valencies (across) and increasing ionic radii (down). The region of cations likely to produce oxides and hydroxides is confined between the dashed lines.

To the left of this region are the cations of strong metals with an 8-electron outer shell, i.e., alkalis and, to some extent, alkali earths (with ionic potentials below 2.0). We know from chemistry that these ions are easily retained and transported in aqueous solutions. In nature, they precipitate from solution only as salts of various acids. The only exception are strongly polarising cations with an 18-electron shell (Cu^{1+} and Pb^{2+}) occurring in nature in the form of oxides.

Cations with very high ionic potentials (over 10), small ionic radii and high charges. All of them are CO_3^{2-}, SO_4^{2-}, PO_4^{3-} etc., which, combining with corresponding cations, give various salts capable of precipitating from solutions.

Among the oxide and hydroxide-forming cations with which we are concerned Mg^{2+}, Fe^{2+}, Ni^{2+}, Zn^{2+}, and Cu^{2+} are easily transported in acid solutions, while in strongly alkaline solutions they precipitate as crystalline hydroxides or basic and neutral salts. The cations with higher ionic potentials, Al^{3+} Fe^{3+}, Mn^{4+}, Si^{4+}, Ti^{4+}, and Sn^{4+} precipitate readily from slightly alkaline or slightly acid solutions as a result of hydrolysis of the salts, mostly in the form of poorly soluble hydroxides (combined with OH anions).

The bulk of hydroxides forms in the oxidised zone of ore deposits and in the *crust of weathering* of rocks in general. Because the majority of hydroxides are poorly soluble in water, in the course of intense oxidation processes they are capable of giving strongly supersaturated solutions. It is natural, therefore, that they are usually found in the form of cryptocrystalline or colloidal masses.

Metal hydroxides (principally of iron, manganese, silicon) are also widespread in *bodies of water* such as marshes, lakes, and seas. Thus, in many contemporary fresh-water lakes in the northern regions (Karelia, Finland, Sweden, Canada) accumulations of iron and manganese hydroxides occur in the shallow near-shore zone as scattered concretions of different sizes and shapes: globular, ellipsoidal, flattened, and various irregular forms. The concretions almost always contain, along with Fe and Mn hydroxides, humic matter, and sometimes Ni, Co, etc.

Irrespective of their mode of formation, hydroxides, especially in air-dry conditions, with time lose their capillary and adsorbed water, and form compounds chemically associated with hydroxyls, and even anhydrous oxides (Fe_2O_3 and MnO_2), especially in regions with a sharply continental climate. In the course of regional metamorphism at moderate depths hydroxides are also converted to crystalline granular masses of anhydrous oxides.

Turning to the question which elements in general form simple anhydrous oxides in the course of *endogenous* processes of the formation of minerals (magmatic, pneumatolitic, and hydro thermal), their list matches exactly that of the cations that tend to form water-insoluble hydroxides in the process of hydrolysis of salts. Such are, for instance, quartz, rutile (TiO_2) cassiterite (SnO_2), corundum (Al_2O_3), haematite (Fe_2O_3), braunite (Mn_2O_3), and many others. The divalent cations adjoining this principal cationic group occur much more seldom as simple anhydrous oxides, but characteristically rather often occur as binary oxides (minerals of the spinel group, the so-called titanates, and tantaloniobates). If we mention in conclusion gaseous carbon and sulphur oxides (CO_2 and SO_2) as well as water (H_2O), this will exhaust the list of elements relating to the class of minerals under consideration.

Peculiar Features of Crystal Structure: Almost all the compounds have crystal structures characterised by ionic bonding of the structural units.

The crystal structures contain the following anions: O^{2-} (in oxides) and OH^{1-} (in hydroxides). The ionic radii of the two are approximately equal (about 1.36 A). Consequently, the entire variety of crystal structures depends in the main on the radii of the cations, their valencies, and the chemical bonds between ions.

In the crystal structures of these compounds, the cations are always surrounded by oxygen or hydroxyl anions, and the coordination numbers of the crystal structures are an important characteristic of these minerals.

Comparing the known structures of *simple oxides* we shall see the different variants of the coordination numbers, from high-coordination ionic structures to the rather rare molecular ones with low coordination numbers and van der Waals bonding. The oxides of divalent metals, characterised by typically ionic bonding, crystallise in the NaCl-type structure, i.e., with coordination numbers 6 and 6. Only the oxides of strongly polarising ions with an 18-electron outer shell have structures of lower coordination, e.g., ZnO (4 and 4) and Cu_2O (4 and 2). The crystal structures of oxides of tri- and tetravalent metals, whose cations are smaller, have lower coordination numbers, which decrease as ionic bonding approaches the covalent type: Al_2O_3 (6 and 4), UO_2 (8 and 4), TiO_2 (6 and 3), SiO_2 (4 and 2). In compounds with molecular structures, the coordination numbers are lower still: 3 and 2 for Sb_2O_3 (cenarmontite), 2 and 1 for CO_2 (solid carbon dioxide).

As to *complex oxides* which comprise metal cations of different sizes, the coordination numbers of each such compound may be either the same or different. For instance, in $FeTiO_3$ (ilmenite), both cations, Fe^{2+} and Ti^{4+}, are surrounded by six oxygen anions. A different picture is presented by compounds of the perovskite ($CaTiO_3$) type: the Ti^{4+}, Nb^{5+}, and other cations are surrounded by the same six anions, whereas the coordination number of the Ca^{2+} and Na^{1+} cations, whose ionic radii are larger, is 12. In compounds of the spinel type ($MgAl_2O_4$) X-ray investigation establishes the coordination numbers of 4 for Mg^{2+}, and 6 for Al^{3+}.

Hydroxides containing hydroxyl groups $(OH)^-$, for instance $Mg[OH]_2$, and also oxides containing hydrogen H^+ as the cation, for instance $HAlO_2$, differ materially in structure from typical oxides. It should be noted that replacement of O^{2-}ions by dipole anions $[OH]^{1-}$ results in the formation of typical layered structures with ionic bonding within the layers and van der Waals bonding between them. This diminishes the symmetry of the crystalline lattice. MgO, for instance, crystallises in the cubic system of the NaCl type, whereas $Mg[OH]_2$ does so in a hexagonal layered structure.

In the same way, Al_2O_3 crystallises in the trigonal system, and $Al[OH]_3$ in the monoclinic, etc. Strongly polarising cations of the cupro type do not form hydrates of their own, but only enter as constituents into the composition of complex salts.

Principal Cations and their Coordination

The crystal structures of oxides are reflected in the physical properties of the minerals. Compounds characterised by ionic bonding possess stable crystal structures, which on the whole are much stronger than in the case of halogens and sulphides. This is an expression of the strong chemical affinity with the oxygen of the metals forming such oxides. The strength of the crystal structures is reflected in the high hardness of these oxides (6, 7, 8 and 9 on the Mohs scale), high chemical stability, infusibility, very low solubility, etc.

Hydrates possessing layered crystal structures are much less stable structurally due to the weak bonding between the layers. They are notable for their ability to split into thin folia along their basal cleavage. The hydrates of divalent metals have low hardness; when these ionic groups are replaced by cations of trivalent metals, hardness increases, especially in the presence of $[OH]^{2-}$ ionic groups in the structures (e.g., diaspore).

The colouring of minerals also has its peculiarities. The compounds in which ions of the noble-gas type (Mg^{2+}, Al^{3+}, etc.) take part are, as a rule, either colourless or allochromatic. However, the vast majority of minerals in which the cations are asymmetrical ions (Fe, Mn, Cr, etc.) are of intense dark colours, black being most prevalent. Many of them are opaque or translucent in thin fragments or polished sections, having predominantly brown or red tinges. This also accounts for their sub-metallic lustre. Their magnetic properties are likewise definitely more pronounced.

Classification of minerals: It is customary to divide all the minerals under this section into (1) anhydrous oxides and (2) hydroxides and oxides containing hydroxyl and hydrogen ions. This classification has been adopted in the present course since it fully agrees with the principles of crystal chemistry.

Either category, besides simple compounds, includes binary or more complex compounds, that formerly were treated as separate classes, such as binary oxides of $RO.R_2O_3$ type, the so-called titanates, nio-bates, and tantalates, i.e., "salts" of hypothetical titanic, niobic, and tantalic acids. X-ray investigation of these compounds shows that their crystal structures have nothing in common with the typical salts of oxygen acids. Far from that, they reveal features definitely placing them with the oxides.

Thus this section should include two classes of compounds:

Class 1. Simple and complex oxides.

Class 2- Hydroxides or hydroxyl-containing oxides.

Simple and Complex Oxides

Included in this class are both simple and complex oxides without hydroxyl ions. There is no point in setting the complex oxides apart, since their crystal structures are either analogous or very similar to those of the simple oxides.

These minerals have mostly rather simple crystal structures. The only group having more complex structures, and therefore standing somewhat apart, is that of quartz SiO_2.

The oxygen cation-to-anion ratios in these minerals may range from 2 : 1 (A_2X) to 1 : 2 (AX_2). Furthermore complex oxides may have different cation ratios, usually 1 : 1 and 1 : 2. Only some rare compounds exhibit more complex ratios.

ICE GROUP

As known, water may be in three states: solid (ice and snow); liquid (rain, mineral springs, rivers, lakes, seas, and oceans); and gaseous (water vapour in the atmosphere and in volcanic exhalations). For its properties water stands apart from oxides of metals and metalloids. Water plays a tremendous part in the chemical processes. Occurring in the earth's crust, since chemical reactions take place mostly in aqueous solutions. Without water, just as without oxygen, no organic life could exist on the Earth.

We shall review this compound here only in its solid and liquid states as the most common in nature.

Ice, H_2O.

Chemical Composition: H 11.2 per cent, O 88.8 per cent. May contain gaseous and solid mechanical impurities.

System, hexagonal; *symmetry,* dihexagonal pyramidal. Crystal structure. Molecular, close to that of the diamond (each H_2O molecule is in four fold coordination). The H_2O molecule itself has a peculiar structure, viz., H^1 protons, being ions without electrons of their own and possessing infinitesimal dimension, deeply embed themselves within the oxygen ion 143. The oxygen atomic nucleus is accordingly somewhat displaced away from the protons. Therefore, although the H_2O molecule is electrically neutral, it carries weak residual charges (positive in the area of the protons and negative on the opposite side). This is why in the structure of ice the weakly bonded H_2O molecules are oriented with their positively charged areas towards the negatively charged areas of other H_2O molecules. As for the general arrangement of molecules, the structure of ice is analogous to that of wurtzite (cf. 96) with the only difference that the places of Zn and S are occupied by H_2O molecules. In such structure (with a coordination number as low as 4) the structural units are far from being closely packed (the spacings are too large), on account of which the specific gravity of ice is lower than that of water.

Habit: The crystals of snowflakes with their hexagonal symmetry ex-hibit a great variety of hexactinal forms of growth. Well-known are the dendrites and lace- (circumscribed).

Pattern Ice Formations: In ice caves, ice crystals occur in regular hexagonal plates, tabular individuals, and intergrowths of complex forms. N.N. Stulov has described ice crystals of unique size and clear-cut faces (up to 40 cm long and up to 15 cm wide) found in permafrost in mine workings in Northeastern Asia. In one case, they were found in the cavities of a thick zone of crushing at a depth of 55 to 60 m where the temperature of enclosing perpetually frozen rocks was 3 to 4°. According to the readings of a contact goniometer, the predominant crystal faces were the hexagonal dipyramid and the pina-coid. In another case, very large columnar ice crystals were found in abandoned mine workings in the oxidised zone of a sulphide deposit. The workings were filled with a solid mass of ice. The ice had cavities with mineralised water and gases under high pressure. The largest ice crystals were 60 cm long and 15 cm across, and occurred as hexagonal prisms truncated by the faces of hexagonal dipyramids.

Aggregates: Often composes massive aggregates of crystalline grains (compact snow and firn in glaciated regions). Glacier ice is composed of very large irregularly-shaped crystalline grains. Familiar to everyone are the stalactitic icicles forming from supercooled water, i.e., from melted snow on the shady side of roofs, as well as in ice caves (stalactites and stalagmites). In the form of hail ice often has a concentrically-banded structure. On a cold autumn morning the ground is often coated with ice efflorescence (hoar-frost).

Colour: Colourless or slightly bluish when in large masses. *Lustre,* vitreous. Optically positive. Refractive indices as low as $n_x = 1.310$ and $n_y = 1.309$. *Hardness,* 1.5. Brittle. *Cleavage,* none. *Specific gravity,* 0.917 (less than that of water). A decrease in specific gravity on crystallisation is characteristic also of metallic bismuth.

Genesis: Formed on the surface of water basins on the

cooling of the water. The slush that is formed first freezes into a floating crust onto which crystalline individuals extending vertically parallel to the axes of sixfold symmetry grow from below. Snow is formed from water vapour in the cold regions of the atmosphere. Under the same conditions cooled objects are coated with hoar-frost. In low-temperature ice caves, water percolating from the surface through cracks and fissures turns into ice.

In the permafrost regions with their severe climate and long winters with little snow, the so-called icings are formed during the cold season. These are enormous sheets of ice spreading at times over vast areas. River icings develop when the river freezes to the bottom in shallow places. Running water seeks an escape to the surface and soaks the snow blanket. In other instances, icings are formed by ground waters emerging from below the permafrost layer in the form of non-freezing springs.

Deposits of ice are well known to everyone. Notable among the ice caves in the regions of cold long winter and short summer is the *Kun-gur Cave* (Perm Region) which attracts numerous tourists by the beauty of crystals glittering on the walls and the ceiling. Another famous ice cave is at *Dobsina* (Slovakia) where over 7,000 sq m are covered with ice. The total volume of ice in the cave is 120,000 cu m, and the ice walls rise to the height of 15 m.

Uses: Ice is used for refrigeration and for different other industrial and domestic applications. In very cold regions where in winter there are no sources of potable water, the latter is obtained by melting river and lake ice.

Water, H_2O

According to chemical composition and genesis, natural waters may be divided into the following principal varieties: (1) sea water containing dissolved chlorides and sulphates of Na, Mg, etc. (about 3.5 per cent); (2) fresh river and lake water; (3) surface rain and ground waters; (4) underground waters;

(5) the waters of mineral springs classified according to their mineral content as carbonated, sulphurated, chalybeate, and other waters. Underground waters rich in calcareous salts are known as hard, in contrast to those poor in these salts, called soft.

Physical and Chemical Properties. Pure water has the highest specific gravity of 1 at 4°. Water vapour pressure relative to air is 0.62. The critical temperature of pure H_2O is 374°, and the critical pressure is 217kg/sq cm. The presence of poorly volatile dissolved constituents greatly reduces vapour pressure and raises the critical point.

Water is an environment for and a most important agent in the chemical reactions that take place in the earth's crust. Its dissolving capacity is a fundamental factor in the transportation of chemical compounds of elements either in true or colloidal solutions (sols) in the course of endogenous and exogenous processes of mineral formation. Its evaporation and temperature drop are the controlling factors in the precipitation of crystalline sediments from saturated solutions. In a number of instances crystallisation water enters into the composition of crystallohydrates in definite stoichiometric ratios.

The electrolytic dissociation of water into H^{1+} and OH^{1-} ions plays an essential part in the processes of hydrolysis of dissolved salts with consequent formation of hydroxides, basic and acid salts.

Water is a most important constituent of hydrogels, being adsorbed on the surface of finely dispersed colloidal particles.

Water, just as carbon dioxide, is a mineralising agent in all metamorphic processes. Even an insignificant water content in solid environments facilitates the processes of recrystallisa tion of minerals.

Genesis: In the course of *exogenous* processes, both at the earth's surface and in the atmosphere, enormous masses of water are in constant circulation, evaporating from the surface of seas, oceans, and especially through vegetation, into the atmosphere and returning as atmospheric precipitation. The surface or so-called vadose waters include soil, ground, artesian, and underground waters as well as many fresh-water and mineral springs.

Great quantities of water also take part in the *endogenous* processes of mineral formation. During volcanic eruptions, large masses of superheated water vapour escape into the atmosphere. The crystallisation of magmas deep in the interior of the earth also releases tremendous amounts of water saturated with diverse dissolved compounds. These so-called juvenile waters are probably the source of some hot mineral springs (in regions of recent volcanic activity, such as California).

Uses: The enormous importance of water in the life of the organic world and in human activity is well known. Thus, to support plant life, soils must contain from 40 to 60 per cent of moisture. The human body contains 70 per cent water and man has to consume on the average 2.5 to 3 litres of water a day. Water resources are used extensively by man for most diverse technical purposes (for steam and hydraulic power plants, irrigation, ore concentration, chemical industry, navigation, etc.). Many mineral springs possess curative properties. Examples are the carbonated mineral springs of Narzan, Karlovy Vary, Baden, etc.; the chalybeate waters of Zheleznovodsk; the sulphureted waters of Matsesta, etc.

Cuprite Group

Cuprite or cuprous oxide (Cu_2O) is the only known metal oxide of A_2O type to occur naturally. An artificial product crystallising in the cuprite structure is Ag_2O which has never been found in nature.

Cuprite, Cu_2O

From the Latin "cuprum "meaning copper. Synonym: ruby copper ore. A brick-coloured ore (due to an admixture of iron hydroxides) and pitchy copper ore (due to admixture of silica and iron hydroxides) are essentially colloidal mineral mixtures.

Chemical Composition. Cu 88.8 per cent. Native copper is often found as mechanical impurity, and in cryptocrystalline varieties, Fe_2O_3, SO_2, and H_2O.

System, cubic; *symmetry* hexatetrahedral. Habit, octahedral, less often cubic or dodecahedral. Crystals usually small.

Occasionally fine slender needles or fibres are found. Occurs more often in massive granular, sometimes earthy (mixed with impurities), aggregates.

Colour: Red to lead-grey (in fine-granular or cryptocrys talline aggregates). Streak, brown-red or dark brownish red (when rubbed against another plate turns yellow). *Lustre,* on fracture, adamantine or *Cuprite* sub-metallic. Thin fragments are translucent, $n = 2.85$ crystal. *Hardness,* 3.5 to 4. Brittle. *Cleavage,* {111} distinct. *Specific gravity,* 5.85 to 6.15.

Diagnostic Features: Identifiable by adamantine lustre, red streak, and especially by paragenetic association with native copper, occasionally with other secondary copper minerals—malachite and azurite. From cinnabar, proustite, and pyrargyrite differs in streak (bright red for cinnabar, red turning green on rubbing for proustite and pyrargyrite, and brown-red for cuprite), but mostly by behaviour before the blowpipe.

In the blowpipe flame on charcoal turns black, then fuses quietly, and in the reducing flame yields a copper globule. When held in pincers, colours the flame light green, and when moistened with HCl, imparts pleasant blue tinge to the flame. Readily dissolves in nitric acid, colouring the solution green. On addition of excess of ammonia the solution turns blue.

Genesis: Cuprite is formed almost exclusively through *exogenous* oxidation of chalcocite and less often of bornite ores, occurring in the zone of secondary sulphide enrichment of copper ore deposits (below the ground-water table).

Massive development of cuprite is liable to occur when for

some reason or other (e.g., lowering of the base level of erosion) the ground-water table sinks and the previously formed chalcocite-rich zone comes within the oxidation zone. Oxidation of chalcocite in the course of dissolution of the evolving sulphuric acid in water may be expressed as follows:

$$Cu_2S + 2O_2 + H_2O - Cu_2O + H_2SO_4$$

It will be easily seen that in the case of oxygen deficiency, the reaction would yield metallic copper, either instead of or in addition to Cu_2O, which is often actually observed in specimens of cuprite (recognised by the hackle appearance of fracture).

Occasionally cuprite occurs together with native copper in certain sedimentary rocks containing vegetal remains. In this case it is in all probability the product of the reduction of cuprous sulphate by organic matter in the presence of some oxygen according to the following reaction:

$$Cu_2SO_4 + C + O - Cu_2O + CO_2 + SO_2$$

In solutions strongly saturated with carbon dioxide cuprite becomes unstable. Pseudomorphs of malachite, more rarely of azurite, after cuprite are quite common.

In the USSR, cuprite occurs in quantity and in well-formed crystals at *Gumeshevskoye, Mednorudyanskoye,* and *Turya* mines in the Urals, at a number of localities in the Altai, and elsewhere, associated with malachite, azurite, iron hydroxides, etc. A notable foreign deposit is at *Chessy* near Lyons (France) where crystals up to 2-3 cm across have been found.

Uses: Cuprite is one of the best ores of copper. In the remote past, the cuprite ores along with native copper must have been worked intensely. At present large cuprite bodies are comparatively rare.

Incite Group

This group includes relatively simple oxides of the *A X* type, i.e., of divalent metals (Mg, Ni, Fe, Mn, Cd, Ca, Be, Zn) as well as Cu, Pb, Pd. We shall discuss here only zincite and tenorite.

Zincite, ZnO

Zn 80.3 per cent, O 19.7 per cent. System hexagonal; symmetry, dihexagonal dipyramidal. Crystal structure of the wurtzite type. Occurs as impregnations and massive.

Colour: Orange-yellow to dark red. *Streak,* orange-yellow. *Lustre,* adamantine. Optically positive. $n_x = 2.029$ and $n_v = 2.013$. *Hardness,* 4. *Cleavage* {1010} distinct. *Specific gravity,* 5.66.

Infusible in the blowpipe flame. Soluble in acids. Has rectifying properties.

Is found in quantity at the well-known contact-metasomatic deposit at *Franklin,* New Jersey (U.S.A.) in calcite masses associated with willemite, Zn_2SiO_4, and franklinite, (Zn, Mn)Fe_2O_4. Crystals are very rare, and are observed only in more recent calcite veins. Finds also recorded from the Olkusz (Poland) lead-zinc deposit, Saravezza (Tuscany, Italy), and other localities.

Tenorite, CuO: Cu 79.9 per cent, O 20.1 per cent. Synonym: melaconite (massive variety).

System, monoclinic. A rare mineral. Commonly occurs in fine-scaly or earthy aggregates.

Colour: Black to grey-black. *Streak,* grey-black. *Lustre,* sub-metallic. Strongly anisotropic in polished sections. *Hardness,* 3.5. Brittle. *Specific gravity,* 5.8 to 6.4. Infusible in the blowpipe flame. Readily soluble in acids.

Occurs in the oxidised zones of copper-sulphide deposits in association with cuprite, limonite, chrysocolla, malachite, manganese hydroxides, and other hypergene minerals. In the USSR, occurs in the area of the Turya copper mines (Northern Urals), Mednorudyanskoye deposit near Nizhny Tagil. Found in quantity in the copper deposits near Lake Superior, Michigan (U.S.A.), and the Atacama desert, Chile. Occurs as thin scales on lava at the Vesuvius and Etna, associated with alkali and copper chlorides.

Corundum-ilmenite Group

Besides dimorphic oxides of the A_2X_3 type (Al_2O_3 and Fe_2O_3), this group, according to X-ray data, includes complex (binary) oxides of ABX_3 type (where A = Mg, Fe^-, Mn^- and B = Ti) crystallising in the corundum-type structure. It should be noted that at high temperatures Fe_2O_3 and Fe^-TiO_3 form a continuous series of solid solutions which disintegrate on cooling.

All the minerals of the group crystallise in the trigonal system and have crystal structures of the same Al_2O_3 type 145. Only Fe_2O_3 occurs in nature in two modifications, viz., a-Fe_2O_3 in the trigonal system, and y-Fe_2O_3 in the cubic system. The symmetry of the $FeTiO_3$-type binary oxides is of a lower order as compared with simple oxides of the Fe_2O_3 type, since their cations are not equivalent. They are formed at relatively high temperatures.

Corundum, Al_2O_3

The name of the mineral comes from India. For Al_2O_3, the following polymorphous modifications are known: (1) a-Al_2O_3 (corundum), trigonal and the most stable in nature; formed over a wide range of temperatures (500 to 1500°); (2) Al_2O_3, hexagonal, stable at very high temperatures; Al_2O_3 occurs at temperatures from 1500 to 1800°; this modification forms in the course of very slow cooling of the Al_2O_3 melt; (3) y-Al_2O_3, cubic with the crystal structure of the spinel type (as in the case of maghemite); it is obtained artificially by ignition of aluminium hydroxide (hoemite) up to 950°; at higher temperatures it becomes unstable and alters to a-Al_2O_3.

Chemical Composition: Al 53.2 per cent. Crystal varieties are of very pure composition. Traces of Cr render it red, of Fe^{2+}—brown (mixed with Mn) and pink, of Ti—blue, and a mixture of Fe^- and Fe^{2+} — black.

System, trigonal; *symmetry*, ditrigonal scalenohedral. Although it seems complex, the cell structure is rather simple in fact. The oxygen ions are in closest hexagonal packing with their sheets perpendicular to the three-fold axis and superimposed on each other. The Al cations are located between two such sheets, in the form of hexagons (with unoccupied centre), and fill two-thirds of the octahedral spaces (i.e., spaces between the six oxygen anions, three of which belong to one oxygen-ion sheet, and three to the other, the latter anions being turned through 180° in respect of the first three). The groups of three oxygen ions each form a common face shared by two adjacent octahedra in the neighbouring sheets. It should be noted that the oxygen sheets are superimposed on each other so that in each column of octahedra two occupied octahedra alternate with an unoccupied one, the pairs of occupied octahedra forming spiral threefold axes in a vertical plane.

Habit: Common are well-formed barrel-shaped, columnar, pyramidal. In a hexagonal cell six Al-O layers are located one on top of another. Most common forms of faces are the hexagonal prism {1120}; hexagonal dipyramids {2241}, {2243}; rhombohedron {1011}; and pinacoid {0001}. The faces of the prisms, dipyramids, and pinacoids commonly have oblique striations. Sometimes the striations are horizontal due to pinacoidal twinning. Corundum usually occurs disseminated in rock, but in some deposits it is granular massive, Colour, commonly bluish- or yellowish-grey (in translucent varieties). Occurs in transparent crystals of different colour. Transparent gem quality varieties are: colourless *leucosapphire,* blue *sapphire,* red *ruby,* yellow *oriental topaz,* violet *oriental amethyst,* green *oriental emerald,* and *asteriated corundum* showing a six-rayed star on the basal pinacoidal plane when the crystal is turned which is due to oriented mechanical microinclusions.

Lustre, vitreous. n_y = 1.767 and n_z = - 1.759. *Hardness,* 9. *Cleavage,* practically none. Displays only parting after pinacoid, sometimes after the unit rhombohedron (due to polysynthetic twinning). *Specific gravity,* 3.95 to 4.10. The melting point of artificial corundum is 2040°.

Diagnostic Features: Readily identifiable by crystal form, striations, often by characteristic bluish-grey colour and high hardness. Differs from similar disthene (Al_2SiO_5) by the absence of perfect cleavage and by high hardness. Ruby differs from red spinel by crystal form. In irregular grains can be distinguished only under the microscope. Infusible in the blowpipe flame. Insoluble in acids.

Genesis and Occurrence: Occurs in alumina-rich and silica-poor plutonic *magnetic* rocks such as corundum-syenites, and anorthosites, associated with feldspars; more rarely in other rocks (andesites and basalts). Corundum-bearing syenite *pegmatites* of economic importance are also known.

Contact-metasomatic corundum deposits in which the gem varieties (ruby and sapphire) are usually found are formed in crystalline limestones in the vicinity of igneous rocks. Corundum deposits sometimes develop by vigorous action of pneumatolytic agents upon alumina-bearing sedimentary and igneous rocks. In such cases, corundum associates with such minerals as andalusite, sillimanite, and also rutile, diaspore, etc.

Regional metamorphism of alumina-rich sediments (bauxites), without any direct relation to igneous rocks, may also give rise to corundum-bearing rocks. In the course of these processes the rocks are usually converted to crystalline schists.

Being very inert chemically, corundum often turns up in *placers.* In the course of hydrothermal processes the earlier formed corundum is sometimes hydrated into diaspore ($HAlO_2$). This phenomenon never occurs near the surface, with very rare and doubtful exceptions.

In the USSR, a most notable location for high-grade corundum is *Semiz-Bugu* in the Bayan-Aul district of Kazakhstan, where it occurs in pocket-like bodies which are almost solid granular dark-blue or grey corundum, separated from surrounding quartzites by andalusite zones. The corundum is paragenetically associated with muscovite, haematite, rutile, and other minerals. A number of corundum deposits are also known on the eastern slope of the Urals, namely in the Kysh-tym district, in the upper reaches of the *Borzovka river,* where it occurs as plagioclase corundum veins in ultrabasic rocks; on the eastern shore of the Irtyash Lake where metamorphic rocks, such as marbles, enclose lenticular and irregular bodies of emery containing chloritoids and sulphides.

Famous foreign localities are the deposits of transparent gem corun-dums (ruby and sapphire) in India, *Upper Burma* (Myanmar) (in the contact zone of marbles and granites), Thailand (in placers, mostly sapphire), and others.

Uses: Due to its high hardness, the chief application of corundum is as an abrasive. Made into wheels, disks, emery papers and powders, it is employed for polishing and grinding (chiefly in metal working). The transparent coloured varieties are used as gemstones.

Of late, in the countries where electricity is cheap, corundum (alun-dum) is made artificially by electric melting of alumina-rich bauxites. Artificial corundum is superior to natural in purity and coarse-granularity. By melting powdered Al_2O_3 with 2.5 per cent Cr_2O_3 rubies identical to natural ones are obtained, while addition of Ti, and also of Co gives sapphires.

Haematite, Fe_2O_3

From the Greek "hematicos" meaning sanguine. Two natural polymorphs of iron oxide are known: trigonal stable a-Fe_2O_3 and cubic and unstable y-Fe_2O_3 which will be reviewed separately. Synonyms: iron glance, specularite, red iron ore (compact cryptocrystalline variety), micaceous haematite (red pulverulent variety). Pseudomorphs of haematite after magnetite are called martite.

Chemical Composition: Fe 70.0 per cent. May contain isomorphous admixtures of Ti (titanhaematite) and Mg. Contains water in negligible quantities (hydrohaematite, usually in colloform state). Cryptocrystalline compact masses often contain silica and alumina as mechanical impurities.

System, trigonal; *symmetry,* ditrigonal scalenohedral. Crystal structure, analogous to that of corundum Habit. Commonly in platy, rhombohedral, and tabular crystals forming in cavities. Most common forms are rhombohedral (1011), (1014), and pinacoidal {0001}, hexagonal dipyramidal (2243), etc. Owing to polysynthetic twinning along the rhombohedron {1011}, pinacoidal planes, like those of corundum, often show triangular markings, while rhombohe dral planes {1011}, show parallel diagonal striation. Simple twinning is rare, mostly along the rhombohedron as well as along the prism. Occasionally occurs as accumulations of slightly curved platy crystals inter-grown in planes close to the pinacoid (so-called *iron roses).*

Aggregates: Commonly massive cryptocrystalline, or in foliated or scaly aggregates. Large kidney-shaped forms of radial-fibrous structure are called *kidney ore*. Very often finely-dispersed anhydrous iron oxide imparts an intense colour to the surrounding minerals and rocks, as in the case of jaspers (siliceous rocks), red marbles, and red clay shales.

Colour: of crystalline varieties of haematite is iron black to steel grey. Very thin translucent plates transmitting deep-red light. Earthy finely-dispersed varieties bright red. *Streak,* cherry-red. *Lustre,* sub-metallic. Sometimes a bluish tarnish is observed. *Hardness,* 5.5 to 6. Brittle. *Cleavage;* none. Characteristic rough parting on rhombohedron {1011}. *Specific gravity,* 5.0 to 5.2.

Diagnostic Features: Readily distinguished from similar minerals (ilmenite, magnetite, goethite, etc.) by cherry-red streak, high hardness, platy and scaly aggregates, and the absence of magnetic properties.

Infusible in the blowpipe flame. A characteristic feature is that in the reducing flame at high temperature it becomes magnetic (turns into magnetite). Slowly soluble in hydrochloric acid.

Genesis and Occurrence: Haematite is formed under oxidising conditions in deposits and rocks of widely different origin. The temperature of its formation may vary within a wide range, but at high temperatures it is unstable.

1. Occasionally occurs in negligible amounts as a constituent of *igneous,* mostly acid, rocks (granites,

syenites, andesites, etc.). Relatively rare in *pegmatites* where it develops at the hydrothermal stage of mineral formation.

2. In certain *hydrothermal* deposits it is found in large masses in association with quartz, barite, sometimes magnetite, siderite, chlorite, and other minerals. Quite common in later reduction to magnetite. The reverse, however, may be observed at other places, i.e., the transformation of magnetite into haematite (the process of martitisation). These phenomena are evidently connected with changes in the oxidation-reduction potential after the deposition of the minerals from hydrothermal solutions.
3. As *a sublimation product,* usually in small amounts, in crystals and crusts on the walls of craters and in cracks in lavas. Thus, in 1871 a body of haematite about 1 m thick was deposited within 10 days in a fissure on Mount Vesuvius. Most probably it is a decomposition product of iron chloride sublimates.
4. In *the crust of weathering* in hot and arid climate haematite and hydrohaematite are formed as a result of dehydration of earlier iron hydroxides. This irreversible process is easily demonstrated artificially by gradually dehydrating goethite. The two minerals are fairly often associated with aluminium hydroxides— diaspore and boehmite (in ferruginous bright-red bauxites).

Also known are reniform haematite-hydrohaematite formations with concentrical conchoidal structure and hackly fracture (kidney ore). Some of the concentrical zones are composed of radial-fibrous masses of haematite or hydrohaematite, sometimes goethite. The conditions of their formation still remain obscure. They may be the products of dehydration of limonite of similar shape.

It should be noted in conclusion that in a hot climate in the upper zones of magnetite deposits extensive martitisation is often observed, i.e., oxidation of magnetite masses into haematite.

5. In the course of *regional metamorphism* at high temperatures and pressures large masses of haematite are often formed in sedimentary deposits of limonite through the dehydration of the latter. Such is the origin, for instance, of oolitic red iron ores, specularite-bearing schists, and ferruginous quartzites in which bands of quartzite alternate with those of scaly compact haematite. Such quartzites may enclose huge bodies of massive haematite-magnetite ores.

As a higher oxide of iron, haematite is chemically inert in the oxidised zone. Occasionally, however, haematite masses are subject to mechanical weathering (disaggregation). Crystalline haematite varieties are converted into hydroxides extremely rarely, under very specific conditions and on a most limited scale.

The largest deposits of high-grade haematite-magnetite ores in the USSR are at *Krivoi Rog* in Pre-cambrian ferruginous quartzites that are a product of regional metamorphism of originally sedimentary ferruginous deposits repeatedly and steeply folded. An example of hydrothermal formations is the deposit at *Kutim* where the ores occur in Paleozoic dolomites and consist of crystalline masses of coarsely tabular haematite, in places altered to magnetite. Beds of oolitic red iron ores are widespread on the western slope of the Middle Urals — in the *Kusye-Alexandrovsk* and *Pashyisk* districts.

Notable foreign deposits of haematite are those of *Lake Superior* in the USA, where they are found in Pre-cambrian metamorphosed series and at *Minas Geraes,* Brazil. Highly interesting mineralogically is a contact-pneumatolytic deposit on the island of *Elbe.* Excellent iron glance crystals from this locality are exhibited in many mineralogical museums.

Uses: Haematite ores are a most important ore smelted for iron and steel. The iron content of massive haematite ores usually varies from 50 to 65 per cent. Pure varieties of pulverulent haematite are used as pigments and in the manufacture of red pencils.

Ilmenite, $FeTiO_3$ or $FeO_;TiO_2$

Named after the Ilmen mountains, Southern Urals, where it was first discovered. Synonym: titanic iron ore.

Chemical Composition: Fe 36.8 per cent, Ti 31.6 per cent, O 31.6 per cent. May contain isomorphous admixture of Mg, sometimes in large amounts (picroil-menite), occasionally Mn (up to a few per cent). Forms continuous isomorphous $FeTiO_3$—$MgTiO_3$ and, probably $FeTiO_3$ —$MnTiO_3$, series.

System, trigonal; *symmetry,* rhombohedral. Crystal structure is analogous to that of corundum with the difference that the sites of Al are alternately occupied by Fe^{2+} and Ti^{4+}. Such replacement by different ions results in lowering the symmetry of the crystal lattice.

Habit: thick tabular, rhombohedral, sometimes lamellar. The most common forms are pinacoid {0001}, rhombphedrons {1011}, {0221}, {2243}, etc. Twins are on rhombohedron {1011}. Commonly occurs as irregular impregnated grains, rarely granular massive. Under the microscope ilmenite, in the form of platy segregations is found in some haematite varieties, as a product of disintegration of solid solutions. Much more often, however, it is found in the so-called titanomagnetites, occasionally in titanious augites, and in other minerals, also as a product of disintegration of solid solutions.

Colour, iron to steel grey. *Streak,* generally black, sometimes dark brown to brownish red (for varieties containing haematite inclusions). *Lustre,* sub-metallic. Opaque. *Hardness,* 5 to 6. *Cleavage,* on rhombohedron {1011} very imperfect. *Specific gravity,* 4.72. Slightly magnetic.

Diagnostic Features: Resembles haematite. In crystals identifiable by crystal form (there are only rhombohedrons, no hexagonal dipyramidal faces). Massive ilmenite is distinguishable from haematite by streak and weakly magnetic character. Infusible in the blowpipe flame. In the reducing flame becomes definitely magnetic. In powdered form dissolves with difficulty in concentrated hydrochloric acid with separation of titanium oxide. When boiled with tin after fusion with $KHSO_4$, imparts a bluish-violet colouration to the solution, which when diluted with water turns pink (titanium reaction).

Genesis: Occurs as impregnations in basic igneous rocks (gabbros, diabases, pyroxenites, etc.), often associated with magnetite, and also in alkaline rocks. Occasionally occurs in quantity in certain pegmatites (syenitic), in paragenesis with feldspars, biotite, ilmenorutile, etc.

In hydrothermally altered igneous rocks, usually decomposes into the so-called leucoxene. Gases are known of ilmenite decomposing into a mechanical haematite-rutile mixture with the external crystal shape of ilmenite being preserved. In the USSR, some rather large crystals of ilmenite are found ingrown in pegmatites in the *Ilmen Mountains* near Miass, Southern Urals, among nepheline-syenites. Widespread as inclusions in titano-magnetites in many deposits.

Important foreign localities are *Ekersund-Zoggendhal* (Norway), where ilmenite occurs in veins in norites (basic igneous rock), and *Kragero,* from which crystals weighing up to 6 and even 7 kg, have been recorded, and elsewhere.

Uses: In large accumulations ilmenite is a source of titanium ore, used in the form of TiO_2 as white pigment (titanium white with high covering power), in iron alloys, ferrotitanium, with 10 to 15 per cent of Ti, in some special steels, etc. Its heat and corrosion resistance, welding capacity, and low specific gravity make metallic titanium an important material of the aircraft industry.

BBAUNITE GROUP

This group includes oxides of Mn, which may contain Fe and differ materially from the corundum minerals in crystal structure.

Bbaunite, Mn"Mn"—O_3.

The formula is often abridged to Mn_2O_3.

Chemical Composition: MnO 44.8 per cent, MnO_2 55.2 per cent. Often contains up to 8 per cent silica and an excess of manganese oxide against the formula. In addition, braunite often contains iron, sometimes up to 10 per cent. The presence of B and Ba is also detected spectroscopically. The presence of the latter element may be explained by the fact that barite is a common accessory of braunite.

System, tetragonal; *symmetry,* ditetragonal dipyramidal. Structure still obscure. Crystals have octahedral habit; the tetragonal dipyramid, approaching the octahedron at the corners, sometimes occurs in combination with prismatic and pinacoïdal faces. More often found in granular aggregates.

Colour, black. *Streak,* brownish black. *Lustre,* sub-metallic. Opaque. *Hardness,* 6. *Cleavage,* {111} distinct. *Specific* gravity, 4.7 to 5.0. Non-magnetic.

Diagnostic Features: Similar to many black manganese minerals. High hardness and brownish-black streak are characteristic features, but positive identification is possible only in polished sections under the microscope and by X-ray analysis.

Infusible in the blowpipe flame. With borax gives the Mn reaction. Dissolves with difficulty in hydrochloric acid, evolving chlorine and leaving a residue of gelatinous silica. In nitric acid decomposes into MnO and MnO_2.

Is formed under recrystal, ducing conditions, but remains stable within a certain range of the oxidation-reduction potential. In more strongly reducing conditions is replaced by hausmannite, Mn_3O_4. Occurs in some *contact-metasomatic* deposits, and also in *hydro-thermal* veins in association with various manganese and iron minerals, barite, quartz, etc. Large masses occur in *regionally-metamorphosed* sedimentary manganese deposits.

In the oxidation zone, braunite is unstable: gradually oxidising to the highest Mn valency, is converted to psilomelane, and then to pyro-lusite, MnO_2, which is more stable in the presence of oxygen.

In the USSR, large masses occur in a number of metamorphosed sedimentary deposits of Central Kazakhstan: at *Jezdy,* Karsakpai district; in this locality it has also been observed as massive granular aggregates in small hydrothermal veins; at *Karajal,* Atasuy district, etc. At *Sapal,* in the Urals, there is a hydrothermal deposit in limestones, where braunite is associated with hausmannite, haematite, magnetite, jacobsite, and sulphides of Fe, Pb, etc.

Outside the Soviet Union, large metamorphosed deposits of braunite are known in India, Brazil, and South Africa (Postmasburg) and elsewhere.

Uses: Braunite ores are an important ore of manganese which go to make ferromanganese, a flux used in steel smelting. Manganese-poor ores are added to blast furnace charges.

Spinel Group

The spinel minerals should be regarded, according to X-ray data, as binary oxides, not as salts of oxygen acids, i.e., not as aluminates, ferrates, etc. The group abounds in isomorphous Al^{3+}, O^{3+}, and Mn^{3+}; the divalent metals are mainly $Mg^{2+}{}_{r}$ Fe^{2+}, sometimes Zn^{2+}, Mn^{2+}; and occasionally, in smaller quantities Ni^{2+} and Co^{2+}. Characteristically, the divalent ions of Pb, Sr, Ca, and Ba with longer radii and also the univalent ions Na and K never enter into the composition of the minerals of this group. Depending on the combination of the elements listed above there are distinguished many mineral species which have much in common in crystal.

Crystal Structure of Spinel: Basic pattern of octahedra with vertical four fold axis arrangement and conditions of formation (mostly developing at high temperatures and pressures).

The vast majority of these minerals crystallise in the cubic system forming crystals of octahedral habit which persists even in those that crystallise in the tetragonal system. Apart from these stands chryso-beryl, a compound of analogous chemical formula. The ionic radius of Be^{2+} is so small that the compound has an entirely different structure, and crystallises in the orthorhombic system.

The crystal structure of the spinel minerals is rather complex. The oxygen ions are closely packed on the planes parallel to the octahedral faces. The divalent cations (Mg^{2+}, Fe^{2+},etc.) are surrounded by four tetrahedrally located oxygen ions, whereas the trivalent cations (Al^{3+}, Fe^{3+}, Cr^{3+}, etc.) are surrounded by six oxygen ions occupying the octahedral apexes. Each oxygen ion is bonded to one divalent and three trivalent cations.

Thus the structure is characterised by the combination of isometric "structural units" —tetrahedra and octahedra, each apex being shared by a tetrahedron and three octahedrons.

These structural features explain very well such properties of the minerals as optical isotropy, the absence of cleavage, chemical and thermal stability of compounds, high hardness, etc.

Spinel, $MgAl_2O_4$: The origin of the name is unknown. Transparent beautifully coloured varieties (red, pink, green, blue, violet, etc.) are known as gem spinels.

Chemical Composition: MgO 28.2 per cent, Al_2O_3 71.8 per cent. Impurities: Fe_2O_3, which imparts a bottle-green colour (chlorospinel); FeO together with Fe_2O_3, which colours spinel brown or black; sometimes ZnO, MnO, and Cr_2O_3.

System, cubic; *symmetry,* hexoctahedral. Crystal structure is typical and has been described above. Habit. Mostly occurs as intergrown octahedral crystals, usually small, but sometimes as large as 25 cm across. Twins parallel to (111) —planes of the closest packing of oxygen ions. Hence the term "spinel law twinning".

Colour, Colourless varieties are rare. Spinel most often has is typical and has been described above. Habit. Mostly occurs as intergrown octahedral crystals, usually small, but sometimes as large as 25 cm across. Twins parallel to (111) —planes of the closest packing of oxygen ions. Hence the term "spinel law twinning".

Colour, Colourless varieties are rare. Spinel most often has different colouring. *Lustre,* vitreous. Optically isotropic. $n = 1.718$ to 1.75. *Hardness,* 8. The presence of Fe_2O_3 and Cr_2O_3 lowers it to 7.5. to 7. *Cleavage,* {111} indistinct. *Specific gravity,* 3.5 to 3.7 (the lowest for the group). Melting point 2150°.

Diagnostic Features: The most characteristic features are the octahedral habit and high hardness. Under the microscope in polarised light, powdered isotropic spinel is easily distinguished from birefringent varieties of corundum similar to spinel in colour and shape. Differs from the other minerals of the spinel group by having the lowest specific gravity.

Infusible in the blowpipe flame. Insoluble in acids.

Genesis and Occurrence: Spinels occur most frequently in *contact-metasomatic* formations in dolomites and magnesial limestones as the product of the action on the latter of the pneumatolytic agents of the magma at high temperatures. Minerals of the same origin are found in paragenetic association with spinels, viz., garnets, pyroxenes, chlorine-bearing silicates, etc.

Spinel occurs occasionally in *pegmatites* and *magmatic* rocks. Some finds have been recorded from strongly metamorphosed rocks, such as gneisses and schists. At the earth's surface, spinel is perfectly stable and is therefore often found in placers.

In the Soviet Union, individual emerald-green spinel specimens were found in placers in the *Kamenka River* valley, in Kochkar district. In this case, spinel, along with some other gemstones, is evidently the product of weathering of local pegmatites. Blue and violet spinels have been found near *Baikal Lake,* and red varieties in the *Shungan* mines in Ferghana (near the Afghan border). Huge deposits of red gem spinel exist on annually yielded 10,000 carats of gem spinel, mined together with rubies (50,000 carats). No large accumulations in primary rocks are known.

Magnetite, $FeFe_2O_4$

The origin of the name is vague. Thought to be derived from the locality Magnesia, (in Macedonia) or from the name of a shepherd Magnes who by ancient tradition discovered the magnetic stone when the iron ferrule of his stick and the nails of his shoes stuck to it. Synonyms: magnetic iron ore, lodestone.

Chemical Composition: FeO 31 per cent, Fe_2O_3 69 per cent. Fe content 72.4 per cent. Usually rather pure. Varieties are *titanomagnetite* containing TiO_2 (up to a few per cent) which at high temperatures enters into the composition of the mineral as a solid solution, and *chromomagnetite* containing Gr_2O_3 (up to a few per cent). Occasionally varieties rich in MgO (up to 10 per cent), Al_2O_3 (15 per cent), etc., are found. Mention also should be made of a rather rare ferromagnetic iron oxide, *y*- Fe_2O_3, crystallising in the cubic system, and known as maghemite (combination of the words magnetite and haematite).

System, cubic; *symmetry,* hexoctahedral. Crystal structure analogous to that of spinel.

Habit: Often octahedral, more rarely rhombic-dodecahedral crystals. Faces {110} often striated parallel to the longer rhombic diagonal. Minute dendrites are observed in basalt glass under the microscope. Twins along (111).

Aggregates: Mostly netite crystal massive granular or impregnating igneous, predomi- *q* {HO}.

Colour, iron black, sometimes crystals show a blue tarnish. *Streak,* black. *Lustre,* sub-metallic. Opaque.

Hardness, 5.5 to 6. Brittle. *Cleavage,* none. Sometimes parting on {111} is observed. *Specific gravity,* 4.9 to 5.2. Strongly magnetic, sometimes showing polarity. At red heat (at about 580°) magnetism disappears abruptly, and reappears on cooling.

Diagnostic Features: Strong magnetic character and black streak usually readily distinguish it from externally similar minerals (haematite, goethite, hausmannite, chromite, etc.), but sometimes it cannot be distinguished from other, rarer minerals species of the spinel group, rich in ferric and ferrous oxides: ferrochromite, jacobsite, etc.

Infusible in the blowpipe flame. In the reducing flame first alters to maghemite (y-Fe_2O_3) and then to haematite, becoming non-magnetic. With borax and salt of phosphorus gives the iron reaction (bottle-green bead). Powdered magnetite is soluble in hydrochloric acid.

Genesis and Occurrence: Magnetite is formed under more strongly reducing conditions than haematite and occurs in ore deposits and rocks of most divers genetic types.

1. In *magmatic* rocks it is usually observed as an impregnation. Magmatic deposits of titanomagnetite in the form of irregular masses and veins are frequently associated genetically with basic rocks (gab-bro).
2. In negligible amounts occurs in many *pegmatites* in paragenetic association with biotite, sphene, apatite, and other minerals.
3. In *contact-metasomatic* formations it is often an important component and is accompanied by garnets, pyroxenes, chlorites, sulphides, calcite, and other minerals. Large deposits have been found at the contact of limestones with granites and syenites.
4. As an accessory mineral, magnetite occurs in *hydrothermal* deposits, mostly in association with sulphides (pyrrhotite, pyrite, chalcopyrite, etc.). Comparatively rarely it forms deposits in its own right in association with sulphides, apatites, and other minerals. The largest deposits of this type in the Soviet Union are located in the Angara-Ilim region in Siberia.
5. *Exogenous* formation of magnetite could have taken place only under exceptional conditions. The presence

of magnetite grains in recent marine silts is believed to be due not only to denudation of dry land but also to new formations from iron hydroxides under the reducing influence of decaying organic matter.

6. In course of *regional metamorphism,* magnetite, as well as haematite, is formed through the dehydration of iron hydroxides (formed in sedimentary rocks exogenously) only under reducing conditions (oxygen deficiency). Of this type are the numerous large beds of haematite-magnetite ores occurring in metamorphosed sedimentary strata.

In the oxidised zone, magnetite is rather stable. On physical weathering of rocks it is freed of the accessory minerals and tends to concentrate in placers. Magnetite is the principal constituent in the black sand residue of panning of gold-bearing sand and gravel.

In conditions of weathering magnetite is highly resistant to hydration, i.e., to conversion into iron hydroxides. The process is observed but rarely and on a comparatively small scale.

The phenomenon of martitisation (formation of pseudomorphs of haematite after magnetite) is observed in hot climatic zones. Some local martitisation of magnetite is also observed in hydrothermal and metamorphosed deposits where it is entirely independent of exogenous processes.

We shall mention but a few of the many magnetite deposits in the Soviet Union.

1. An example of magnetic deposits is the *Kusinskoye* titanomagnetite deposit, enriched with vanadium. The deposit (18 km north of Zlatoust in the Urals) comprises veins of solid ore in altered parent igneous rocks of the gabbro formation. Magnetite is closely associated here with ilmenite and chlorite.
2. An example of metasomatic deposits is the well-known *Mount Magnitnaya* (Southern Urals) where thick bodies

of magnetite occur in garnet, pyroxene-garnet, and other skarns formed through the action of granitic magma emanations on limestones. In some portions of the ore bodies, magnetite is associated with primary haematite. The ores beneath the oxidised zone contain disseminated sulphides (pyrite, occasionally chalcopyrite, galena, etc.).

Of the same type are the deposits of *Mount Vysokaya* (near Nizhny Tagil), *Mount Blagodat* (Kuvshino district), a cluster of deposits in the Kustanai Region, and also *Dashkesan* in Azerbaijan, etc.

3. The great deposit at *KrivoiRog*, the Ukraine, belongs to the *regional-metamorphosed* sedimentary type. Besides typical beds, the ferruginous quartzite series contains massive columnar iron ore bodies of a lenticular cross-section persisting to very great depth.

Of similar origin are the deposits of the so-called *Kursk magnetic anomaly* (southeast of the city of Kursk); the magnetite-haematite ores in Central Kazakhstan, etc.

Notable foreign localities are *Kirunaoaara* and *Luossavaara* (Sweden), where the ores occur in thick vein-like bodies in metamorphosed strata where the magnetite is associated with apatite. Well-known are the great magnetite-haematite deposits in the *Lake Superior* district, (U.S.A.) in very ancient metamorphosed schists, in Labrador (Newfoundland), and elsewhere.

Uses: Magnetite ores often containing up to 60 per cent Fe are a most important raw material for smelting of iron. Undesirable impurities are phosphorus (maximum permissible content for the Bessemer process is 0.05 per cent, and only 0.03 per cent for high-grade steels) and sulphur (maximum content 1.5 per cent). In the Thomas process, in which the phosphorus is removed with the slag, its content should not be below 0.61 per cent and not over 1.5 per cent. The phosphorus slag from this process is used as a fertilizer.

The slag obtained in smelting titanomagnetite ores is a source of vanadium which is a valuable alloying metal for high-grade steels. Vanadium pentoxide is used in the chemical industry as a pigment in ceramics and in other applications.

Chrome-spinellids with the common formula (Mg, Fe) (Cr, Al, Fe)$_2$O$_4$. All the mineral species classified under this heading occur in the same conditions and are so similar in external features that species of different composition cannot be distinguished without chemical analyses. In the prospecting and mining practice they are known under the general name of *chromites.* According to composition the following principal mineral species are recognised: *chromite* proper, ($FeCr_2O_4$) (occurring in meteorites, and very rarely in the earth's crust); magno-chromite, (Mg, Fe)Cr_2O_4; alumochromite, Fe(Cr,Al)$_2$O$_4$; and chromopicotite, (Mg,Fe)(Cr,Al)$_2$O$_4$. Chromite was first mentioned by Academician Lovits as early as 1797.

Chemical Composition: For the most common chrome-spinellids, the Cr_2O_3 content varies within a wide range— from 18 to 62 per cent; that of FeO from 0 to 18 per cent; MgO from 6 to 16 per cent; Al_2O_3 from 0 to 33 per cent; Fe_2O_3 from 2 to 30 per cent. Moreover, there may be isomorphous impurities such as TiO_2, up to 2 per cent; V_2O_3, up to 0.2 per cent; MnO, up to 1 per cent; ZnO, up to a few per cent; NiO, up to tenths, and CoO, up to hundredths of one per cent.

System, cubic; *symmetry,* hexoctahedral. Crystal structure analogous to that of spinel.

Habit: Occasionally small octahedral crystals, but usually rounded or somewhat irregular grains, or massive granular aggregates.

Colour, black. In polished sections, semi-transparent or deep red to brown-red when viewed in transmitted light. Only varieties rich in FeO and Fe_2O_3 are absolutely opaque. *Streak,* brown. *Lustre,* sub-metallic, $n = 2.07$ to 2.16. *Hardness,* 5.5 to 7.5. *Cleavage,* none. *Specific gravity,* 4.0 to 4.8. FeO- and Fe_2O_3-bearing chrome-spinellids show weak magnetism, while varieties rich in these components and poor in Cr_2O_3 are strongly magnetic.

Diagnostic Features: The common distinguishing features of the chrome-spinellids are black colour, brown streak, high hardness, and chrome reaction. The minerals occur so constantly in ultrabasic rocks (dunites, peridotites, and serpentinites) that in the field these features alone are sufficient for almost positive identification.

Infusible in the blowpipe flame. The borax or phosphate bead is emerald-green when cold (chrome reaction). Insoluble in acids.

Genesis and Occurrence: Almost exclusively confined to *magmatic* ultrabasic rocks both as impregnations and as massive accumulations mostly of irregular pocket-like, lenticular, and columnar shapes. Constantly associated with chrome-spinellids are greenish serpentine (hydrosilicate of Mg and Fe), olivine, $(Mg, Fe)_2SiO_4$, chromium-bearing chlorites, sometimes chromous emerald-green garnets, etc. In some ultrabasic rock masses, chrome-spinellids may be paragenetically associated with platinum and osmiridium minerals.

In the weathering zone the chrome-spinellids are chemically inert. Only in hot climate they are subject to oxidation and decomposition. In the Soviet Union, most of chrome-spinellid ore deposits are confined to the Ural Mountains. Among them are the deposits of high-grade chromite ores on the eastern slope of the southernmost part of the Urals ridge known as the *Donskoye* or *Kempirsaiskoye* (Aktyubinsk Region). Another important locality is *Saransk* in the Northern Urals (6 km from Biser Station) comprised of steep vein-like bodies of massive chromite ores of a rather poor grade (high iron content). Smaller deposits are found in Transcaucasia: at *Shorjin, Geydarin,* and elsewhere (on the eastern shore of Lake Sevan and southeast of it in Armenia).

Important foreign localities are in *Rhodesia* (South-East Africa), *New Caledonia, Turkey, Cuba,* and elsewhere.

Uses: Chromite ores are the only source of chrome which is of great value as an alloy metal for making special steels. Moreover, it is extensively used as a coating for protecting various metal articles and machine parts from corrosion. Chromite ores are also used as a material for making stable paints, tanning agents, and certain chemicals (potassium dichromates, etc.). Massive low-grade ores poor in Cr_2O_3 and rich in FeO and Fe_2O_3 go into refractory brick.

Hausmannite, $MnMn_2O_4$.

Chemical Composition: MnO 62.0 per cent, MnO_2 38.0 per cent (Mn 72.0 per cent). Also contains FeO and Fe_2O_3.

System, tetragonal; *symmetry,* ditetragonal dipyramidal. Crystal structure, the same as that of spinel, though the mineral crystallises in the tetragonal system. Usually massive granular. Crystals occur only in cavities. Habit close to octahedral.

Colour, black. *Streak,* brown to reddish brown. *Lustre* of unoxidised crystals strongly adamantine or sub-metallic. *Hardness,* 5. Brittle. *Cleavage,* {001} perfect. *Specific gravity,* 4.7 to 4.9. non-magnetic.

Diagnostic Features: In granular aggregates is distingui shed with difficulty from other manganese minerals without the aid of the microscope. Differs from braunite by lower hardness and obviously reddish streak. Pyrolusite, MnO_2, has a black streak. The psilomelane minerals have either a black or chocolate-brown streak. Differs from haematite in hardness and structure of aggregates.

Infusible in the blowpipe flame. Borax bead turns violet in the oxidising flame. Dissolves in hydrochloric acid with evolution of chlorine.

Genesis and Occurrence: Like magnetite, hausmannite is formed in a more strongly reducing environment than braunite. Like braunite, with which it is often associated, it occurs in Certain *contact-metasomatic* and *hydrothermal* manganese deposits. Minerals rich in manganous oxide, MnO, such as tephroite, Mn_2SiO_4, manganosite, MnO, rhodo-chrosite, $MnCO_3$, manganous garnets, etc., are usually observed in close association with hausmannite. The most frequent non-metalliferous associate is barite, not quartz, the latter being the common associate of braunite and manganese silicates (rhodonite and bustamite).

In large masses hausmannite together with braunite, jacobsite, magnetite, and other anhydrous oxides of Mn and Fe, is widespread in metamorphosed sedimentary manganese deposits. In conditions of weak regional metamorphism, hausmannite derives from the dehydration of manganese hydroxides and also through reduction from pyrolusite and braunite. Hausmannite pseudomorphs after braunite have been observed.

In the Soviet Union, hausmannite is one of the main ore minerals in the *Sapalskoye* hydrothermal deposit in marbled limestones (Nizhny Tagil). Besides hausmannite, the ores contain braunite, magnetite (probably manganous), haematite, rhodochrosite, iron sulphides, ferruginous chlorites, and other minerals. In Central Kazakhstan, hausmannite occurs in association with braunite in the ores of the *Karajal* metamorphosed sedimentary deposit.

Outside the Soviet Union hausmannite occurs in approximately the same paragenetic associations in contact-pneumatolytic deposits at *Nordmark* and *Langban* (Sweden), in vein deposits at *Ilefeld* near Hartz (Germany), and elsewhere.

Uses: Hausmannite ores, as well as those of braunite, are used for making ferromanganese or as an addition to blast-furnace charges (low manganese — ores).

Gahnite, $ZnAl_2O_4$

Chemical composition ZnO 44.3 per cent, Al_2O_3 55

Impurities: MgO, FeO, MnO, and Fe_2O_3.

System, cubic; *symmetry* hexoctahedral.

Habit: The same as that of spinel. Also occurs in rhombic dodecahedrons and distorted cubes. Sometimes individual crystals attain large dimensions. Twinning according to the spinel law, on (111),

Colour, dark-green, greyish-green to black-green. *Streak,* grey. Translucent. *Lustre,* vitreous, n = 1.78 to 1.82. *Hardness,* 7.5 to 8. Brittle. *Cleavage,* {111} indistinct. *Specific gravity,* 4.0 to 4.6. Resembles many other spinel minerals.

Diagnostic Features: Gahnite may be tentatively distinguished from other spinels by higher specific gravity, refractive index, and paragenetic association with zinc minerals, but positive identification is possible only by blowpipe and chemical analysis.

Infusible in the blowpipe flame. With soda on charcoal yields a ZnO coating. Insoluble in acids and alkalis.

Genesis and Occurrence: Occurs comparatively seldom in certain pegmatites and contact-metasomatic deposits in marbled limestones together with other minerals of the same origin, including zinc minerals.

Occasional finds are recorded from strongly altered rocks, such as talc schists.

In the Soviet Union gahnite is found at Altyn-Tau (northwestern part of the Kara-Kalpak Autonomous Republic), in the Ukraine, in pegmatites on the western shore of the Sea of Azov, etc.

It is found in commercial quantity in the well-known *Franklin* deposit in New Jersey (U.S.A.) noted for the unusual association of manganese and zinc minerals. There gahnite is associated with zincite, franklinite, and other minerals. It was found also in the talc schists of Falun (Sweden) and in the diamond placers of Brazil.

Franklinite, (Zn, Mn)Fe_2O_4

System, cubic; occasionally contains FeO and Mn_2O_3. In many physical properties resembles magnetite. Opaque. *Streak* reddish-brown. Slightly magnetic. *Specific gravity,* 5.07 to 5.22.

$n = 2.36$. Infusible. With borax in the oxidising flame yields a reddish-violet bead (manganese reaction). On charcoal with soda produces a ZnO coating. Soluble in hydrochloric acid. If Mn_2O_3 is present, chlorine is evolved from the solution.

Forms thick beds in the well-known contact-metasomatic deposit at Franklin, New Jersey (U.S.A.). Associated with it in crystalline limestones are zincite, willemite (Zn_2SiO_4), calcite, less often gahnite, axinite, manganese silicates, apatite, and other minerals.

Chrysoberyl, $BeAl_2O_4$

Chrysos is the Greek for gold. Varieties: the emerald-green gemstone *alexandrite* (violet-red when viewed under incandescent light).

Chemical Composition: BeO 19.8 per cent, Al_2O_3 80.2 per cent. Always contains impurities: Fe_2O_3(3.5 to 6 per cent), sometimes TiO_2 (up to 3 per cent) and Cr_2O_3 (up to 0.4 per cent) responsible for the emerald-green colour.

System, orthorhombic; *symmetry,* rhombic dipyramidal. Crystal structure. Already Y:S. Fyodorov, when he was developing his system of classification of crystals according to external form, established that structurally chrysoberyl should be similar to the minerals of the olivine group. X-ray investigations have fully corroborated his conclusion that the arrangement of ions in the chrysoberyl unit cell is perfectly analogous to that in the unit cell of forsterite. The oxygen ions actually occupy positions of closest hexagonal packing; the beryllium ions, just as the ions of silicon in forsterite, are surrounded by four oxygen ions, while the aluminium ions are surrounded by six.

Habit: Tabular, sometimes short-or long-prismatic. The closest almost hexagonal packing of the basic oxygen ions results in a pseudohexagonal ratio between the *a* and *b* axes; hence, the frequent pseudohexagonal habit of the crystals both with regard to angles and faces. This is especially marked in (030) trillings characteristic of alexandrite. The (001) faces are often striated parallel to the *a* axis, which is a sure sign of the trilling intergrowth. Chrysoberyl has not been observed in granular masses.

Colour, usually yellow or greenish-yellow, rarely colourless. Only the chrome-bearing variety (alexandrite) is emerald-green (violet-red in incandescent light). *Lustre*, vitreous, greasy on fracture. n_x = 1.753, n_y = 1.747, and n_z = 1.744. *Hardness*, 8.5. Brittle. *Cleavage*, distinct parallel to (110) and indistinct parallel to (010), (001). Fracture conchoidal. *Specific gravity*, 3.50 to 3.84.

Diagnostic Features: Occurs exclusively in crystals of very characteristic shape, sharply differing from that of beryl, high hardness, and specific colour.

Infusible in the blowpipe flame. Insoluble in acids. Decomposes only when fluxed in powdered form with KOH or $KHSO_4$. On ignition the melt moistened with cobalt nitrate shows a specific blue colouring.

Genesis and Occurrence: This rather rare mineral occurs usually in *pegmatites* and *contact-pneumatolytic* formations at the boundary between schists and granite intrusives. In the oxidised zone it is very stable. Occurs in placers, sometimes as rounded pebbles. Some alexandrite crystals are several centimetres across.

The most remarkable group ever found comprising 22 large crystals and small trillings of alexandrite is on view at the Mineralogical Museum of the USSR Academy of Sciences. Chrysoberyl is commonly associated with emerald, feldspars, apatite, and other minerals of mica, chlorite, and talc schists, usually at the contacts of granite pegmatites. Notable foreign deposits are at *Minos Geraes* (Brazil) where chrysoberyl, associated with topaz, rock crystal, spinel, garnet, tourmaline, and other minerals, occurs in pegmatite veins in gneisses and mica schists; and also in placers on *Ceylon* and *Madagascar*.

Uses: Beautifully coloured transparent varieties are used as gemstones.

Rutile Group

This group comprises compounds of the $-4X_2$-type crystallising in the tetragonal system: binary oxides of Ti, Sn, Mn, and Pb. Of these, TiO_2 occurs in three different polymorphous modifications which have special names, and MnO_2 at least in two. There is some evidence that SnO_2 too is dimorphous.

The principal members of the group are not associated paragenetically, and are formed under different conditions. Although occurring under conditions similar to those under which TiO_2 is formed, SnO_2 behaves in the geological processes quite differently. MnO_2 and PbO_2 are formed under entirely different conditions, almost exclusively in the course of exogenous processes of mineral formation.

The notable feature of the chemical composition of rutile (as well as of brookite and anatase) is the presence of Nb^{5+} and Ta^{5+} together with Fe^{2+} as isomorphous impurities.

Since the length of the radii of all these ions is approximately of the same order, the crystal structure of the minerals generally follows the same pattern, but its dimensions increase corresponding to the ionic radii of Nb^{5+}, Ta^{5+}, Fe^{2+} (or Mn^{2+}). Thus, these varieties of rutile are isomorphous mixtures of TiO_2, or rather Ti_3O_6 with $Fe^{-} Nb_2O_6$ or $Fe–Ta_2O_6$, i.e., with the minerals mossite and tapiolite crystallising in the same rutile structure.

It should be noted that $FeNb_2O_6$ and $FeTa_2O_6$ are dimorphous and crystallise both in the tetragonal and orthorhombic systems. The crystal structure of the orthorhombic modification of columbite and tantalite is also similar to that of the orthorhombic modification of brookite (TiO_2).

Rutile, TiO_2

From the Latin *rutilus,* meaning reddish. It is the most stable modification of TiO_2, both at high and low temperatures.

Chemical Composition: Ti 60 per cent. Chemical analyses show frequent presence of other elements: Fe (as ferrous or ferric oxide), sometimes Sn^{4+} (up to 1.5 per cent), rarely Cr^{3+}, V^{3+}, and some others. A variety rich in $FeTiO_3$ (in the form of solid solution) is called *nigrine.*

System, tetragonal; *symmetry,* tetragonal dipyramidal. Crystal structure is typical of the entire group. It has some peculiar features. Whereas in the corundum-type structure the sheets of most closely packed oxygen ions are perpendicular to the three-fold axis, and in the spinel-type structure they are parallel to the octahedron faces (i.e., also to the three-fold axis), in the rutile-type structure (as has been shown by N.V. Belov) the closest packing directions in the form of columns are parallel to the principal (fourfold) axis of rutile crystals. Each Ti ion is surrounded by six oxygen ions occupying the corners of an almost regular octahedron.

In the rutile crystal structure, the octahedra are elongated parallel to the *c* axis in the form of rectilinear columns; hence the acicular or rod-like habit of rutile crystals with cleavage planes parallel to the elongation. The rutile structure differs characteristically from other TiO_2 modifications in that each TiO_6 octahedron shares two edges with adjacent octahedra. Since the columns of closely packed oxygen ions elongated parallel to the rutile four fold axis are in the main oriented in one of the three possible directions in the plane of the hexagonal closest packing, it is not accidental that the rutile group minerals occur as geniculated penetration twins and trillings sometimes at an angle close to 120° (i.e., corresponding to the directions of a hexagonal network). Even ring-shaped sixlings may form in this manner. This is also the cause of ordered (at 120°) growth of extremely thin needles or tiny prismatic crystals of rutile on the basal {0001} pinacoid of haematite, mica, and other minerals, with faces corresponding to the planes of the closest-packed oxygen ions.

Habit: Is very characteristic: prismatic, columnar to acicular. Common forms are {100}, {110}, {101}, {111}, occasionally {001}.

Often striated parallel to the principal *c* axis. Geniculated twins with a (Oil) twinning plane are very frequent. Flat reticulated twins of acicular rutile are called sagenite. As has been stated above, ordered intergrowths of tiny crystals of rutile with haematite crystals are common, the four-fold axis of rutile coinciding with a horizontal two fold axis of haematite. Puffs of acicular capillary crystals of rutile are sometimes found occluded in transparent quartz crystals.

Colour, generally dark yellow, brown, red, and black (nigrine). Colourless or faintly coloured varieties are extremely rare. *Streak,* yellow to light-brown. *Lustre,* adamantine to sub-metallic (black opaque varieties). $n_x = 2.903$; $n_y = 2.616$. *Hardness,* 6. Brittle. *Cleavage,* {110} perfect and {100} distinct. *Specific gravity,* 4.2 to 4.3.

Diagnostic Features: Tetragonal prismatic crystals and geniculated twins are very characteristic. May be confused with minerals of similar habit: zircon ($ZrSiO_4$) which has a higher hardness (7 to 8), cassiterite (SnOg) which has a higher specific gravity. Capillary rutile crystals may be mistaken for tourmaline, from which it differs in optical constants.

Infusible and unchangeable in the blowpipe flame. Insoluble in acids, gives the titanium reaction with salt of phosphorous (in the reducing flame yields a violet head).

Genesis and Occurrence: In nature rutile is formed under a wide variety of conditions. Occasionally it is a constituent of *igneous rocks* (syenites and less often granites). In small quantities occurs in *pegmatites* and certain *hydrothermal* deposits where it is associated with quartz, minerals of titanium and iron (ilmenite, ilmenorutile, haematite, and magnetite), sometimes with corundum, silicates, and other minerals.

Rare finds of neogenesis of rutile have been recorded from the *exogenous* products of decomposition of titanium minerals, occasionally in sedimentary rocks, and in bauxite deposits. Most often, however, rutile is formed in the course of *metamorphic* processes as a result of alteration of titanium minerals, segregating as individual grains in gneisses, mica schists, amphibolites, and other rocks.

Very spectacular are the acicular and capillary crystals in veins of the *Alpine cleft type,* often occluded in rock crystal and haematite crystals; it is often accompanied by brookite and anatase.

In the oxidised zone rutile is chemically inert and therefore often found in placers as rounded grains and pebbles.

In the Soviet Union large crystals of rutile have been found in a number of deposits in mica schists, in the pegmatite veins of the *Ilmen Mountains,* at *Semiz-Bugu* (Central Kazakhstan), where it is associated with corundum, and elsewhere. Moreover, it often occurs in placers, especially in the Middle Urals.

A notable foreign location is in North Caroline (U.S.A.) where excellent rutile crystals of diverse habit are found.

Uses: A source of ferrotitanium for making steels with high impact strength; as a brown pigment for ceramics; as a radio-wave detector; for the manufacture of titanium white, etc.

Brookite, TiO_2

System orthorhombic; *symmetry* rhombic dipyramidal. Crystal structure is projected on plane (100), which is the plane of the closest hexagonal packing of oxygen ions. If the lower drawing is superimposed on the upper one, it will be easily seen that in the general direction of the *a* axis we have a combination of hexagonal and cubic packings (topaz packing); the oxygen ions "75" do not match with the "25"ions, as it should be in the closest hexagonal double-layer packing, but follow the pattern of the cubic closest packing.

The Ti ions lie between the sheets of oxygen ions and have a six fold coordination. Also in agreement with this structure is the flattened (on 100) habit of crystals, four faces of the prism {021} parallel to the *a* axis and the lower drawing is superimposed on the upper pinacoids {001}' forming an almost regular hexagon in cross-section. No less characteristic is the vertical striation of the faces.

Colour, yellow- or red-brown to black. *Streak,* colourless to greyish-or brownish-yellow. *Lustre,* adamantine. Some specimens show sharply different refractive indices with a corresponding change in the dispersion of optical axes for different wavelengths. *Hardness,* 5 to 6. *Cleavage,* {110} imperfect. *Specific gravity,* 3.9 to 4.0 (lower than that of rutile but higher than in anatase). On ignition specific gravity increases and becomes equal to that of rutile (probably a rearrangement of the crystal structure takes place).

Occurrence: Excellent brookite crystals are found in the Atlyanskaya gold placer near Miass, and in Alpine clefts elsewhere in the Urals.

Anatase, TiO_2

System, tetragonal; *symmetry,* ditetragonal dipy-ramidal. Crystal structure is charone, it will be easily seen that in the general di-rection of the *a* axis we have a combination of hexagonal and cubic packings (topaz packing); the oxygen ions "75" do not match with the "25"ions, as characterised by the closest cubic packing of oxygen ions, with a vertical four-fold axis. Although the co-ordination numbers are the same as for rutile (6:3), the geo, it should be in the closest hexagonal double-layer packing, but follow the pattern of the cubic closest entry of the coordination is distorted, because the TiO_2 octahedra are joined in such a way that they have four common edges.

Crystals are of characteristic dipyramidal habit the (Oil) dipyramid being more acute than the (012) dipyramid which is close to an octahedron in shape. More rarely, anatase occurs in prismatic and tabular crystals.

Colour, brown to black. *Streak,* colourless. *Lustre,* adamantine. n_y = 2.55 and n_z = 2.49. *Hardness,* 5 to 6. *Cleavage,* perfect parallel to (001) and (Oil), in which it differs from rutile. *Specific gravity,* 3.9 (lower than that of rutile and brookite).

Infusible in the blowpipe flame. Occurs in pegmatites and chlorite and mica schists. Well-developed crystals are fairly often observed on quartz in Alpine clefts in the Alps, Brazil, and in the Northern Urals. The mineral is chemically inert, occurs in placers such as Atlyanskaya placer in the Southern Urals near Miass and elsewhere.

Cassitebite, SnO_2

Kassiteros is the Greek for tin. Synonym: tin stone. The only economically important ore of tin packing. The Ti ions lie between the sheets of oxygen ions and have a sixfold coordination.

Chemical Composition: Sn 78.8 per cent (according to the chemical formula). Almost invariably contains impurities. In many cases, especially when occurring in pegmatites, contains Fe_2O_3, Ta_2O_5, Nb_2O_5, TiO_2, MnO, FeO, occasionally ZrO_2 and WO_3. All these metals of different valences are probably present as isomorphous admixtures or the products of disintegration of solid solutions, as in the case of certain varieties of rutile.

Crystal structure, identical with that of rutile.

Habit: Well-developed crystals of Cassiterite are frequently found in cavities. Crystals are usually small but sometimes may be as large as 10 cm (weighing several kilograms). Mostly of dipyramidal, pyramidal prismatic or columnar habit, sometimes acicular. In pegmatites the most common are crystals of dipyramidal habit, often twinned. The grains impregnated through greisens and granites are often irregular. The same is true of the Cassiterite grains formed metasomatically in the course of endogenous oxidation of stannite and other sulphurous compounds of tin. Twins are very frequent and, like those of rutile, geniculated.

Aggregates: Granular masses are rare. Usually occurs as tiny impregnated crystals or irregular grains. In the cavities of hydrothermal veins, druses of well-formed crystals are found occasionally. The so-called wood tin occurs in nodules and sinters of concentrically-banded structure characteristic of colloidal accumulations, under the microclosest packing. In contrast to rutile structure, these octahedra share three common edges. Also in agreement with this structure is the flattened (on 100) habit of crystals, four faces of the prism {021} parallel to the *a* axis and the scope it can be often seen in these cases that some zones are composed of cryptocrystalline aggregates, and some of distinctly granular, columnar, radially-oriented individuals.

Colour, because of the presence of Fe, Nb, Ta, and Mn impurities, cassiterite is usually dark-brown to pitch-black. Thin polished sections often show a banded structure of individual crystals and grains due to the alternation of zones of different colour intensity. Transparent varieties are extremely rare. *Streak* of dark varieties is usually of light-brown. *Lustre,* adamantine, on fracture pitchy, slightly greasy. Crystal faces are sometimes dull. Opaque black varieties may even have sub-metallic lustre. $n_x = 2.09$ and $n_y = 1.99$. *Hardness,* 6 to 7. Brittle. *Cleavage,* {110} imperfect, sometimes distinct. Fracture often conchoidal. *Specific gravity,* 6.8 to 7.0. Non-magnetic, though iron-enriched black varieties are electromagnetic.

Diagnostic Features: In crystal form, twinning, and colour resembles rutile, and in light-coloured varieties also zircon. Differs fundamentally from them in specific gravity, hardness (7 to 8 for zircon), and characteristic slightly greasy or pitchy lustre on fracture. In thin sections, small grains of cassiterite may be mistaken for zircon whose birefringence is much lower.

Infusible in the blowpipe flame, but strongly ignited with three volumes of soda on charcoal, in the reducing flame, gives small malleable globules of tin and a white coating of SnO_2. Insoluble in acids. If a drop of hydrochloric acid is placed on cassiterite and touched with a piece of zinc (or better with a specially made zinc needle), after a while, under the reducing effect of intensely evolving hydrogen, a metallic tin coating will appear on cassiterite. This coating becomes shiny when rubbed against a piece of cloth (a very characteristic and almost always successful test for cassiterite).

Genesis and Occurrence: Gassiterite deposits are genetically associated with acid igneous rocks, mostly granites. In granites proper, however, cassiterite occurs very rarely, and when it does, it is in greisenised sections, i.e., those transformed by pneumatolytic agents (F, Cl, B, etc.) into mica-feldspar-quartz rock with topaz, fluorite, lepidolite (lithium mica), tourmaline, and other minerals. Presumably, at high temperatures tin is transported in the form of volatile SnF_4 and $SnCl_4$ compounds, which are later on hydrolised producing an SnO_2 precipitate. It has also been established that alkaline solutions containing hydrogen sulphide vigorously transport tin in a reducing environment.

Cassiterite forms most unevenly distributed accumulations in *pegmatite* dikes associated with tin-bearing intrusions. Unlike hydrothermal cassiterite, the cassiterite found in pegmatites often contains Nb,Ta,Fe, and other metals. Found in par agenesis with it are quartz, micas, albite, tourmaline, sometimes columbite, beryl, spodumene, etc. Gassiterite is also found in some *contact-metasomatic* deposits in close asso, elation with different sulphides, which shows that it deposited during the hydrothermal stage.

Hydrothermal cassiterite vein deposits are much more important economically. Of these two types are most important: (1) quartz-cassiterite, and (2) sulphide-cassiterite veins. The first, apart from the predominant quartz and cassiterite, usually contain tourmaline, white mica, feldspars, wolframite, and small amounts of arsenopyrite, pyrite, fluorite, topaz, beryl, and other minerals. Gassiterite mostly occurs as impregnations in a quartz groundmass and in cavities as crystals, which are sometimes very large.

In deposits of the second type, which are of economic importance in a number of regions of the Soviet Union, cassiterite is chiefly associated with sulphides: in some cases mainly with pyrrhotite, and partly with sphalerite, chalcopyrite, and stannite; in other cases, it is predominantly associated with sphalerite and galena; and elsewhere with various rare sulphides, such as bismuthinite (Bolivian type). The non-metalliferous associates of cassiterite, besides quartz, are rather abundant black tourmalines and frequently ferruginous chlorites and carbonates.

Owing to its high chemical inertness in the oxidised zones of tin ore deposits, cassiterite passes into placers.

Exogenous cassiterite formed in the course of disintegration of tin sulphides is found in the oxidised zones as porous and earthy masses.

In the Soviet Union cassiterite deposits are found in Eastern and especially in North-Eastern Siberia.

Important foreign localities are the *Malayan* tin-bearing province (Burma (Myanmar), Western Thailand, the entire Malay Peninsula, the southern islands of Banka and Billiton) with widespread cassiterite placers derived from the weathering of primary pegmatite and quartz-cassiterite veins. *Bolivia* has numerous deposits—quartz veins containing cassiterite, tungsten minerals, various sulphides, fluorite, tourmaline, as well as sulphide-cassiterite deposits.

Uses: Cassiterite is the most important ore of tin accounting for practically all of industrial production of tin. Tin has the following applications: (1) in the manufacture of tin plate; (2) in the preparation of fusible acid-resistant alloys with copper (bronze), with zinc, copper, and lead (brass), with lead (solder), etc.; (3) in tinning copper ware; (4) in the production of tin foil; (5) in ceramics (for enamels and pigments) and for other purposes.

Columbite-tantalite, $(Fe,Mn)Nb_2O_6$-$(Fe, Mn)Ta_2O_6$

Form a continuous series of isomorphous mixtures. The name *columbite* derives from the American name for niobium—columbium. Synonym: nio-bite.

Chemical Composition Varies Greatly: Even in the same deposit the contents of Fe, Mn, as well as of Nb and Ta vary within a wide range. It should be noted that tantalites occur either as almost pure mangano-tantalites or pure ferrotantalites (intermediate varieties are rare). May contain as impurities negligible amounts of Sn (1 to 2 per cent, seldom up to 9 per cent in tantalites), W, and Ti. However, unlike other tantalo-niobates, they never have any admixture of such large cations as Na, Ca, U, Th, rare-earths, and yttrium.

Crystal structure analogous to that of brookite. The ions of Nb, Ta, Mn, Fe, W, and possibly of Sn occupy the same positions as Ti ions in brookite, i.e., the centres of the oxygen octahedral groups. Habit, lamellar on (010), tabular, sometimes short-columnar. The most common forms are pinacoids {100}, {010}, {001}; prism {110}, dipyramid {111}, etc. Twins of (201) are often lamellar-cordate with characteristic pinnate striation.

Colour, black to brownish-black. *Streak,* red or reddish-brown to reddish-black. *Lustre,* sub-metallic. Opaque. *Hardness,* 6. Brittle. *Cleavage,* {100} rather distinct. *Specific gravity,* 5.15 to 8.20, increases with Ta content. Columbite is a conductor of electricity.

Diagnostic Features: Golumbite and tantalite are almost indistinguishable by external features. They may be mistaken for: (1) ilmenite (distinguished by streak, habit, and negative reactions for Nb and Ta); (2) wolframite, which has a more perfect cleavage on {010} and lower hardness; (3) orthite which has a lower specific gravity (3.2 to 4.2) and light streak; (4) samarskite, aeschynite, euxenite, and other nio-bates and tantalates containing rare earths and radioactive elements, from which they cannot be distinguished without spectral and radio-metric analyses, and microchemical tests.

Infusible in the blowpipe flame. Insoluble in acids. When fused with KOH and treated with dilute hydrochloric and sulphuric acids, on addition of metallic zinc, columbite develops a stable blue colouring. Tantalite, when fused with $KHSO_4$ and treated with hydrochloric acid, colours the solution yellow and yields a heavy white precipitate which also turns bright-blue on addition of Zn; however, on dilution with water, the blue colour disappears.

Genesis and Occurrence: Usually occur in *pegmatite* dikes in association with various minerals formed at the later stages of the pegmatitic process: albite, quartz, muscovite, tourmaline, zircon, wolframite, cassiterite, sometimes samarskite, monazite, etc.

Being rather inert in the oxidised zone, they occur in placers.

Important foreign localities are Moss, Kragero, and Finbo (Norway); near Limoges (France), and also at Ivigtut (Greenland) where very well-formed crystals have been found, etc.

Uses: When found in quantity, these minerals are of economic value as a source of niobium and tantalum used in the production of special steels and in other applications.

Pyrolusite, MnO_2

Pyros is the Greek for fire and *lusios* destructive (it is used as a decolouriser of green glass). Synonym: polianite (the name formerly applied to phanerocrystalline varieties).

Chemical Composition: Mn 63.2 per cent. In fine-granular and cryptocrys-talline masses usually contains mechanical impurities such as Fe_2O_3, SiO_2, H_2O, etc.

System, tetragonal; symmetry ditetragonal dipyramidal. Crystal structure, analogous to that of rutile. Very seldom occurs in crystals (only in cavities) of acicular or rod-like habit. Usually found in crystalline or cryptocrystalline, often pulverulent, sooty masses, sometimes as pseudomorphs after reniform masses of psilomelane.

Colour, black, sometimes with a metallic bluish tarnish. *Streak,* black. *Lustre,* sub-metallic. Opaque. *Hardness,* for crystal individuals 5 to 6; for aggregates, decreases to 2 (depending on their porosity and friability) Very brittle. *Cleavage,* {110} perfect, very characteristic of pyrolusite. *Specific gravity,* 4.7 to 5.0.

Diagnostic features. Differs from other black manganese minerals with a black streak by characteristic cleavage, brittleness, and rather low hardness.

Infusible in the blowpipe flame. Giving off some of its oxygen (up to 12 per cent by weight) changes to lower oxides and becomes brown. Does not alter on heating up to 500°C but in the interval between 550 and 650°, according to X-ray analyses, dissociation takes place with formation of /9-braunite (cubic modification); when heated further (940 to 1100°) /3-braunite is converted into hausmannite, which is stable at high temperatures.

Dissolves in hydrochloric acid with evolution of chlorine. Extensive use of this property is made by the chemical industry. With borax and salt of phosphorus in the oxidising flame yields a violet bead which becomes colourless on reduction.

Genesis and Occurrence: Rather rarely formed in hydrothermal manganese deposits, and then only in an obviously oxidising environment. However, it is widely distributed on the earth's surface as a multiple oxide in near-shore facies of *sedimentary* deposits. It is the most stable manganese oxide in the oxidised zone where it is the end product of alteration of all the manganese minerals containing manganese in low degrees of oxidation. Therefore pyrolusite often occurs in pseudomorphs after manganite, vernadite, psilomelane, hausmannite, etc. Owing to its brittleness, very rarely occurs in placers.

Constantly occurs in the so-called manganese hats, i.e., in the oxidised zones as well as in a number of sedimentary deposits. One of the world's largest sedimentary deposits is at *Chiatury* (Georgian Soviet Socialist Republic), where pyrolusite occurs as oolitic nodules or pseudomorphs after manganite oolites in the form of cryptocrystalline soft aggregates (at the outcrops); at *Nikopol* (the Ukraine) where it occurs in larger globular concretions with a concentric-banded structure (usually as pseudomorphs after manganite).

Notable foreign localities for pyrolusite are the oxidised zones of metamorphosed deposits in India, Ghana (West Africa), and elsewhere. Well-formed crystals were found at Flatten (Czechoslovakia).

Uses. Pure pyrolusite ores have a wide variety of applications: (1) in the making of dry batteries; (2) in making activated components for such batteries; (3) as a decolouriser of green glass; (4) in the manufacture of chemicals for medicinal and other purposes; (5) in the manufacture of special gas masks against carbon monoxide, of catalysts of the hopkalite type to purify the exhaust fumes of automotive engines, etc.; (6) in the production of drying and various other oils, wax, as a tanning agent for chrome leather, in photography, in the preparation of pigments, etc. The ore used for the manufacture of dry batteries should contain not less than 80 per cent of manganese dioxide.

Perovskite Group

Perovskite, $CaTiO_3$, is similar in chemical formula to ilmenite which crystallises in a structure typical of corundum, but is very different from ilmenite in crystal structure, since the Ca^{2+} ion has a much longer radius than the Mg^{2+}, Fe^{2+}, and Mn^{2+} ions. The latter feature is also reflected in the association of the elements in the minerals of the perovskite group.

To begin with, isomorphous admixtures to Ca in some minerals are rare earths: Ce, La, and also Yt, which is quite natural (provided that Ca^{2+} is replaced by isometric Na^{+} ions).

Then, as in the rutile group, Ti^{4+} may be partly or largely replaced by Nb^{5+} or Ta^{5+} provided that the summary charge of all cations is not disturbed, i.e., Ca^{2+} should be replaced by univalent cations. Indeed, in such cases, an isometric Na^{+} ion appears as an isomorphous admixture to Ca^{2+}. The crystal structure of the compound remains intact. It should be noted that artificial $NaNbO_3$ has the same crystal structure as perovskite; the places of Ca^{2+} are taken by Na^{+} and those of Ti^{4+} by Nb^{5+} ions. Small amounts of U^{4+} and Th^{4+} may sometimes be present as isomorphous admixtures to Ca.

Perovskite, $CaTiO_3$

Named after L. A. Perovsky.

Chemical Composition: CaO 41.1 per cent, TiO_2 58.9 per cent. May contain in negligible amounts the following impurities: Fe (up to 2 per cent), Cr, Al, and TR (knopite).

System, externally cubic; optically, however, perovskite is usually anisotropic with twinning lattices on faces, a sign of its transformation into a lower-temperature orthorhombic modification. According to N.V. Belov, in the perovskite crystal structure, which follows the closest packing pattern, the calcium cations, being rather large, participate in the closest packing of oxygen anions. On the section parallel to (001), the O^{2-} ions are staggered with Ca^{2+} ions as in the NaCl structure. This is why the crystals of perovskite have cubic shape and cubic cleavage.

Habit: Frequent crystals, which are sometimes rather large, have a cubic habit. The cube faces are striated parallel to the edges which is probably due to twinning in the course of transformation into the orthorhombic modification. Also occurs in reniform masses in which small cubes are discernable.

Colour, greyish black, reddish brown, orange-yellow, and light-yellow. *Streak,* white or greyish-yellow. *Lustre,* adamantine, $n = 2.34$. *Hardness,* 5.5 to 6. *Cleavage,* cubic, distinct. *Specific gravity,* 3.97 to 4.04.

Diagnostic Features: Perovskite is characterised by the cubic habit of its crystals with faces having perpendicularly oriented short striations (parallel to the edges). Differs from minerals of similar colour and hardness by very faintly coloured or white streak.

Infusible in the blowpipe flame. Soluble only in boiling sulphuric acid or upon fusion with $KHSO_4$.

Genesis and Occurrence: Perovskite does not occur in large masses. Is found as impregnations in certain alkali basalts (melilitic, leucitic, etc.), sometimes in titanomagnetite and chromite deposits. Chrome-bearing perovskite has been found in the *Saranovskoye* chromite deposit (the Urals). The best perovskite crystals have been found in contact-metasomatic deposits of unusual origin (associated with basic igneous rocks) in the *Nazyam* and *Shishim* Mountains in the Zlatoust district (the Urals)—in the well-known Akhmatovskaya, Nikolaye-Maximilianovskaya, Melnikovskaya, and other mines, very interesting from the viewpoint of paragenetic mineral associations.

Perovskite occurs in many localities outside the USSR, the most notable of which are those in the Zermatt Valley (Switzerland). Perovskite has no economic importance of its own.

Loparite, (Na,Ce, Ca) (Nb, Ti)O_3

First discovered by L.G. Kuznetsov on the Kola Peninsula.

Chemical Composition, Variable: According to chemical analyses, loparite contains (per cent): TiO_2 39.2 to 40, Tr_2O_3 32 to 34, $(Nb,Ta)_2O_5$ 8 to 10, CaO 4.2 to 5.2, Na_2O 7.8 to 9.0, SrO 2.0 to 3.4, K_2O 0.2 to 0.7, UO_2 hundredth fractions, ThO_2 0.2 to 0.5, SiO_2 0.2 to 0.7; in altered varieties, may contain up to 3.5 per cent of H_2O.

System, cubic; symmetry hexoctahedral. Crystal structure, similar to that of perovskite.

Habit: Mostly in cubic crystals, sometimes with truncated {111} faces, usually very small (1 to 2 mm and occasionally 1.5 cm across). Almost invariably twinned according to the fluorite law.

Colour, black or greyish-black. *Streak,* brown. Lustre, sub-metallic. Thin polished sections, brownish-red in transmitted light, $n = 2.24$. Occasionally shows optical anomalies. *Hardness,* 5.5 to 6. *Cleavage,* none. Fracture uneven. *Specific gravity,* 4.75 to 4.89.

Diagnostic Features: In addition to colour, streak, and hardness, is characterised by penetration twins according to the fluorite law.

Infusible in the blowpipe flame. Powdered loparite fuses with difficulty with soda on charcoal. A solution of the flux in hydrochloric acid boiled with tin turns violet (reaction of titanium reduction), then light blue (niobium reaction). In the presence of tin, the bead of salt of phosphorus turns yellow in the reducing flame; on cooling turns violet (titanium reaction). Insoluble in acids. Completely decomposed by HF. Readily fuses with $KHSO_4$.

Genesis and Occurrence: Loparite is a rock-forming *magmatic* mineral. Occurs as more or less uniformly disseminated crystals in dark aegirite-enriched ($NaFeSi_2O_6$) nepheline-syenites (Lovozero tundra).

Occurs also in many pegmatite dikes as concentrated impregnations or as small crystals in feldspars, eudialyte (zirconium silicate of complex composition), and in other minerals.

Common paragenetic associates of loparite are feldspars (microcline, albite), nepheline, aergirite, eudialyte, sphene, and some other minerals. Loparite crystals, especially large ones, often contain inclusions of the surrounding minerals, which indicates that they have formed metasomatically, i.e., are metacrystals.

Uses: Loparite is economically important as a source of niobium, rare earths, titanium, and other elements. Niobium, in the form of ferroniobium has been extensively used lately in making high-grade steels and heat-resistant alloys with aluminium, nickel, and other metals. When added to steels, ferroniobium imparts to them a number of valuable properties: (1) resistance to air hardening; (2) increased elasticity and flexibility; (3) improved welding capacity and resistance to heat and acids. Extra-hard niobium and tantalum alloys are widely used in the manufacture of drills and saws for cutting hard steels, as well as in making springs for watches. Moreover, niobium and tantalum are used for the manufacture of filaments for special-purpose electric lamps. Tantalum is also used as a substitute of platinum, iridium, etc., in powerful electronic tubes.

Pyrochlore Group

Included in this group are minerals close to the perovskite group in chemical composition but differing from the latter in chemical formulas. Of the numerous minerals of this group we shall describe pyrochlore, aeschynite, and samarskite.

Pyrochlore, (Na, Ca. . .)$_2$(Nb, Ti. . .)$_2$0$_6$ [F, OH]

Greek *pyros* means fire, *chloros*, green (before the blowpipe certain varieties turn yellowish-green).

Uranium and thorium-enriched varieties in the metamict state contain much H_2O (7 to 14 per cent), no alkalis, very little calcium oxide (betafite group). Only the external cubic of octahedral crystal forms, sometimes with curved faces, have been retained.

Chemical Composition: Chemical composition is extremely variable. The content of individual constituents in the pyrochlore-microlite (tantalum analogue) isomorphous series varies as follows (per cent):

> Nb_2O_5....63 to 0; Na_2O....1 to 6; $(Ce, La)_2O_3$....2 to 13.3
>
> Ta_2O_5....0 to 77; K_2O....0 to 1.4; $(Y, Er)_2O_3$....0 to 5.1
>
> TiO_2....2 to 13.5; CaO.....4 to 18.1; SnO_2......0 to 4.0
>
> ThO_2....0 to 5; MnO....0 to 7.7; ZrO_2......0 to 5.7
>
> UO_2....0 to 11.4; FeO.....0 to 10; WO_3......0 to 0.3
>
> UO_3....0 to 15.5; Fe_2O_3....0 to 9.7; H_2O......0 to 6.0

Also present sometimes are small amounts of Sb_2O_5, MgO, PbO (probably as a product of radioactive decay), HfO_2, SiO_2, Al_2O_3, SrO, BeO, CuO, Bi_2O_3, CeO_2, etc.

Thus, the principal variations in the chemical composition of these minerals are expressed in the fluctuations in the content of Nb_2O_5 and Ta_2O_5.

Many minerals, bearing different names, with certain peculiar features of chemical composition and physical properties, are tentatively included in this group, although they have not yet been studied well enough.

System, cubic; *symmetry* hexoctahedral. Crystal structure. Specimens that have not undergone metamict decay (poor in radioactive elements) have cubic crystal structure.

Habit: Octahedral Crystals may be as large as 2 cm across. Octahedral crystals are often truncated by cubic faces. Also massive and colloform. Ordered intergrowths with zircon are observed on (111) planes, the face of the pyrochlore octahedron coinciding with that of the tetragonal zircon {111} dipyramid.

Colour, dark-brown, reddish-brown, yellowish-green, chlore crystal, occasionally brownish-black. Light-yellow rims may develop on the surface (transformation into an optically-anisotropic mineral poorly studied as yet). *Streak,* yellowish-white to light brown or reddish-yellow. *Lustre,* adamantine, greasy, vitreous or resinous (metamict varieties). Semi-transparent or translucent. *Hardness,* 5 to 5.5. Brittle. *Cleavage,* practically none. *Specific gravity,* 4.03 to 4.36. Tantalum varieties have higher specific gravity (up to 4.9). Sometimes strongly radioactive. Altered varieties give a flash on heating (at about 500°).

Diagnostic Features: Is characterised by octahedral habit, colour, greasy adamantine lustre. Another characteristic feature is its association with alkaline igneous rocks (nepheline-syenites and syenite-pegmatites). It is very similar to zircon with which it sometimes forms intergrowths, and to scheelite, from which it differs by hardness and absence of cleavage.

Before the blowpipe fuses with difficulty into a black-brown globule. Certain varieties when heated become yellowish-green. With borax yields a bead which is reddish-yellow in the oxidising and dark-red in the reducing flame. Soluble in strong sulphuric and hydrofluoric acids.

Genesis and Occurrence: Occurs in pegmatites of nepheline-syenites, in association with feldspars, mostly with albite, sometimes with zircon, biotite, apatite, ilmenite, sphene, calcite, and Hardness, 5 to 6. Fracture conchoidal. Specific gravity, 5.16 to 5.23. Infusible but intumescent in the blowpipe flame. Insoluble in acids other than HF.

Occurs in pegmatites of nepheline-syenites in association with nepheline, feldspars, biotite, muscovite, magnetite, zircon, and other minerals. Has been recorded from the mines of the Ilmen Reservation, Hitthero (Norway), and elsewhere.

Samarskite, (Y, Er...)$_4$[(Nb, Ta)$_2$O$_7$]$_3$

System, orthorhombic. Named after Samarsky, Chief of Staff of the Corps of Mining Engineers.

Occurs in prismatic, sometimes lamellar crystals with dull faces which are mostly pinacoids {100}, {010}, and prisms {101}. Also as irregular grains, occasionally massive. Notable are its intergrowths with columbite. Sometimes columbite crystals *Columbite* are overgrown by a thin samarskite coating which faithfully reproduces the columbite habit.

Colour, velvet-black, brownish-black. *Streak,* reddish-brown to black. *Lustre,* characteristically strongly resinous. Almost opaque. Thin polished sections, brown in transmitted light, n = 2.21 to 2.25 (metamict varieties). Sometimes has strong birefringence. *Hardness,* 5 to 6. Fracture *Samarskite*conchoidal. Specific gravity, 5.6 to 5.8. Oriented inter growth of samarskite with columbite. Magnified. Strongly radioactive.

Diagnostic Features: Diagnostic features of samarskite are velvet-black dull crystal faces, conchoidal fracture with strong resinous lustre.

Before the blowpipe fuses slightly at the edges into black glass. In the closed tube decrepitates and becomes incandescent. With salt of phosphorus yields the emerald-green bead of uranium both in the oxidising and the reducing flames. Fuses with $KHSO_4$. Solution in hydrochloric acid acquires a light-blue coloration (niobium reaction) when boiled with tin or zinc.

Occurs in pegmatite dikes of the Ilmen Reservation in the Urals, where it is associated with feldspars, which are tinted reddish-brown near samarskite (owing to the radioactive decay of samarskite elements), as well as with columbite, aeschynite, monazite, garnets, tourmaline, and other minerals. Outside the Soviet Union, occurs in deposits in North Carolina (with columbite) and Maryland (U.S.A.), on Madagaskar (Antanamalasa), and near Moss (Norway).

Often, when closely intergrown with albite, imparts to the latter a characteristic brownish-yellow coloration.

Uraninite Group

The group comprises dioxides of tetravalent metals: U, Th, and Zr. We shall describe only uraninite here.

Uraninite, UO_2

Name derives from composition. The most important source of uranium and radium.

Chemical Composition: Chemical composition of the naturally occurring crystals does not correspond to its written formula, being intermediate between UO_2 and UO_3. The presence of U^{6+} in uraninite is possibly due to oxidation. Contains Ra, Ac, Po, and other products of radioactive decay. Radiogenic Pb is always present in uraninites as the end product of U and Th radioactive decay (in the form of Pb^{206}, Pb^{207}, and Pb^{208} isotopes), its content often being as high as 10 to 20 per cent. Uranium ores, however, often contain ordinary lead (as galena inclusion) which, besides the isotopes mentioned above, also contains Pb^{204} in a constant quantity (up to 10 per cent of the total content of isotopes). Some uraninite varieties called *cleveite* and *nivenite* contain rare earths (Ce, La, Er...) and also Yt (their total reaching 12 per cent). Coarse-crystalline varieties occurring in pegmatites are notable for the Th content. Occasionally, considerable quantities (up to 7.5 per cent) of Zr may be present. Gases that may be occluded in uraninite are He, Ar, N, Co_2, etc.

Helium is invariabiy a product of radioactive decay. Almost

invariably present is H_2O which enters into the composition of the mineral in the course of alteration.

System, cubic; *symmetry* hexoctahedral. Crystal structure is of the fluorite type.

Habit: cubic with rudimentary octahedral and rhombic dodecahedral face development. Occurs in octahedral, occasionally in rhombodo-decahedral crystals. Crystals are usually small (up to 1 cm across), sometimes ingrown in rock. Penetration twins according to the fluorite chlore into a secondary product in the form of pseudo-morphs after octahedral pyrochlore crystals. It should be noted that the pseudomorphs sometimes have hedral crystal of intumescence cracks.

This newly formed mineral is optically isotropic. Colour, light-yellow, Pyrochlore is sometimes found as a neogenetic primary pyrochlore formation in the ilmenite aggregates deriving probably from endogenous decomposition of ilmenorutile (Ti, Nb, Fe)O_2.

Uses: When found in commercial quantities has economic importance as a source of niobium, tantalum, and uranium.

Aeschynite, (Ce, Ca, Th...) (Ti, Nb)$_2$O$_6$

From the Greek *Aeschyne,* shame (in the past chemists did not know how to decipher its composition). It is a titanoniobate of rare earths. With priorite (tantalum analogue) yields isomorphous mixtures (with rare earths and yttrium). The content of principal constituents varies as follows (per cent):

Ce_2O_3.....15.5 to 19.5; TiO_2.....21.2 to 23.9; SnO_2.....0 to 0.2

(Y, Er)$_2O_3$.....0.9 to 4.5; Nb_2O_5.....23.8 to 32.5; MnO.....0 to 0.03

CaO.....2.3 to 2.7; Ta_2O_5.....0 to 6.9; MgO.....0 to 0.01

FeO.....2.2 to 4.3; ThO_2.....11.2 to 17.5

System, orthorhombic. Crystals are prismatic, sometimes up to 12 cm long. Colour, black, fallow. *Streak,* dark brown, almost black. *Lustre,* adamantine, greasy. The varieties that underwent metamict decay are isotropic.

Aggregates: Mostly as colloform and reniform masses. These varieties are called *pitchblende.* Dull sooty coatings or pulverulent masses are known as *uranium black.*

Colour, black, sometimes with a slight violet tinge. Thin polished sections, viewed in transmitted light, opaque or dark-brown and greenish in strongly altered portions. *Streak,* brownish black, slightly shiny. *Lustre,* sub-metallic, mostly typically pitchy, waxy or dull for strongly altered varieties. *Hardness,* 5 to 6, for strongly altered varieties as low as 3. Brittle. Fracture, uneven, nearly conchoidal. *Specific gravity,* 10.3 to 10.6; lower for strongly altered varieties. Usually ranges from 8 to 10 and may be as low as 6.5 and even 4.5. Other properties. Strongly radioactive.

Diagnostic Features: Characteristic black colour, strong pitchy lustre on fracture, high specific gravity, and strong radioactivity. For oxidised specimens, a very characteristic feature is association with brightly orange or yellow products of uraninite or pitchblende alteration.

Infusible in the blowpipe flame. Strongly oxidised uraninites rather readily soluble in HNO_3, H_2SO_4 and HF, but very slowly in HCl. The varieties containing rare earths are the most soluble. The borax bead becomes green in the reducing and yellow in the oxidising flame. Neutralised with ammonia, the solution yields a yellow precipitate of ammonium uranate $(NH_4)_2UO4$.

Genesis and Occurrence. The following principal genetic types of uraninite deposits are known:

1. Accumulations of uranium minerals in granite and syenite *pegmatites* where uraninite is comparatively rare and unevenly distributed in paragenesis with the minerals of thorium, rare earths, niobium, tantalum (struverite, fergusonite, monazite, etc.), as well as with

tourmaline, zircon, feldspars, micas, sometimes in association with uraniferous carbonaceous compounds such as thuch-olite, carburan, and other minerals of indeterminate composition. Uraninite formed at high temperatures contains, as a rule, thorium and rare earths.

2. Uranium oxides occur in far greater quantities in *hydrothermal* deposits, where pitchblende is sometimes associated with arsenides of Ni and Co (niccolite, smaltine, rammelsbergite, chloanthite), native bismuth, sometimes bismuthinite, native arsenic, native silver, and argentite. Pitchblende is in some cases associated with haematite, but more often with calcium and iron carbonates, black decomposed fluorite, and other minerals. However, the bulk of pitchblende occurs independently in carbonaceous veins and stringers in association with rare sulphides, sometimes black fluorite, and other minerals.

At Gilpin, Colorado (U.S.A.), uraninite occurs in quartz-siderite veins as colloform segregations in association with pyrite, chalcopyrite, sphalerite, galena, rare native bismuth, and other minerals. In a gold Guadalupe deposit, at Chihuahua (Mexico) uraninite was found in calcite veins with gold and pyrite.

3. Uranium oxides in the form of uranium black are also formed through *exogenous* weathering of uranium deposits in the lower parts of oxidised zones or in the so-called cementation zone (below the ground-water table) in fissures in ore bodies or the country rocks, and even in the cracks of individual minerals. Uranium black is probably formed by the reduction of water-soluble compounds of hexavalent uranium in an oxygen-deficient environment, compounds being transported by percolating waters to the cementation zone. Uranium sub-oxides possess very low solubility and, therefore, precipitate from solution. Uranium black

occurs as most thin coatings, veinlets, and sooty masses of dark-grey or velvet-black colour, sometimes with a slight brown tinge. Under the same conditions but in wider fissures, more compact uranium oxides may be deposited in layers forming dull black crusts of colloform texture.

Uranium black is also found in some *sedimentary* rocks deposited under reducing conditions, especially in the presence of organic remains. Its frequent associates are impregnated iron sulphides (pyrite and marcasite), phosphates, and other minerals. Such sedimentary deposits are often enriched with molybdenum and vanadium. It should be noted that, despite their low uranium tenor, the bulk of uranium is contained in such deposits.

Uraninite always suffers some degree of later alteration irrespective of the type and age of deposit. The fact that it always contains higher uranium oxides shows that it is highly oxidisable. In the oxidised zones it decomposes much more readily than in endogenous conditions and is therefore a source of numerous exogenous uranium minerals, which (depending on the weathering conditions, the composition of primary ores and of the surface waters) may be hydroxides, sulphates, carbonates, uranovanadates, uranophosphates, and uranosilicates. They are all brightly coloured yellow, green, or orange. Their mode of occurrence shows that uranium compounds in the oxidised zone are subject to migration. Although in this process some portions of the zone are depleted and some enriched with uranium, on the whole a considerable part of it is dispersed.

A great deposit of uraninite lies beyond the Arctic Circle at the *Great Bear Lake* (Canada). It was discovered by its bright yellow-orange outcrops of the secondary uranium minerals (the elevated sections of the area have been scoured clean of drift by glaciers). The numerous ore-bearing carbonaceous and quartz veins, lenses and stringers traced over a vast area contain carbonates, haematite, pitchblende, sulphides and arsenides of cobalt and nickel, native bismuth, and more recent native silver, argentite, pyrite, chalcopyrite, galena, sphalerite, etc. The ore minerals very frequently occur as metacol-loidal aggregates. The deposit is enclosed in pre-cambrian rocks. Its age determined by different methods is 1,300,000,000 years.

Another rich deposit is located in the *Katanga* province of the Congo. It is variously known as Kasolo, Shinkolobwe and Kalongwe. Mineralisation there is in the form of lenses and veins in dolomite rocks and schists (at a considerable distance from granite masses). The ores and wall rocks contain, besides quartz, occasional tourmaline, monazite, apatite, and also sulphides (pyrite, linneite, chalcopyrite) and selenides of nickel, cobalt, etc. The pitchblende which sometimes occurs as large masses is usually recrystallised into coarse-granular aggregates. Uraninite crystals are also common. Diverse secondary uranium minerals such as janthinite, becquerelite, schoepite, sklodowskite, kasolite, soddyite, uranospherite, curite, etc., have been observed here in the oxidised zone in considerable amounts. The age of the deposit is about 610,000,000 years.

Uses: It is interesting to note that until radium was discovered by the Curies uraninite ores extracted as a by-product of silver mining were used only in the manufacture of yellow, orange, and black pigments. At the turn of the century uranium ore wastes after the extraction of radium were also used for this purpose. It was only recently that the problem of release and utilisation of the tremendous nuclear energy was solved, that the great value of these ores was discovered. The USSR was the first country to harness nuclear energy for peaceful purposes by building the world's first atomic power plant.

The amount of radium present in uraninite is as low as hundred thousandths of one per cent of the total uranium mass. Radium preparations obtained by the treatment of uranium ores have many medical applications such as in the treatment of cancer (in the form of radon, a product of radium decay) for which purpose they are placed in special "guns";

the gamma-radiations of radium is used for detecting flaws in metal castings, reinforced-concrete, etc.

Quartz Group

The members comprising this group have the same very simple composition: SiO_2, and are nothing but a series of polymorphous modifications.

In crystal structure they stand completely apart from other oxides. As will be shown later, they are directly related to silicates in crystal structure. Therefore some authors classify this group of minerals with the silicates. However, since quartz is chemically a typical oxide, according to the classification adopted, we should review it in this section.

The three principal polymorphous modifications of SiO_2 have names of their own: quartz, tridymite, and cristobalite.

A characteristic of crystal structure of quartz and of other polymorphous modifications is that the Si^4 ion is always surrounded by *four* O^{2-} ions which occupy the apexes of a tetrahedron. Each apex of a tetrahedron is also the apex of the adjacent one. Thus, the crystal structure of these minerals consists essentially of tetrahedra framework with common apexes. Although the mode of linkage is the same for all the modifications.

Generally speaking, the packing of oxygen ions is not compact, since there are "voids" between the tetrahedra. In the low-temperature modifications they are small and in the high-temperature more "loosely" packed modifications they are larger. This is directly reflected in specific gravity and refractive indices.

Since each oxygen ion is shared by two adjacent SiO_4 tetrahedra, it is located between two silicon ions, whereas each Si^{4+} ion is in fourfold oxygen coordination. Hence the coordination numbers 2 and 4. In each SiO_4 tetrahedron the positive Si^{4+} ion cancels out only half the negative valence of each surrounding O^2 ion, but since each oxygen ion is also shared by another tetrahedron, the second half of its valence will be cancelled out as well. Thus, it will be easily seen that on the whole the chemical formula for the compound should be SiO_2.

a-Quartz, SiO_2

This modification, stable at low temperatures, is usually called simply quartz; the origin of the name is unknown. Quartz is one of the most widespread and best studied minerals in the earth's crust.

Chemical Composition: The written formula probably adequately expresses the composition of colourless transparent varieties. Milk-white and other variously coloured varieties may contain mechanical impurities of gases, liquids, and solids in varying amounts: CO_2, H_2O, hydrocarbons, NaCl, $CaCO_3$, sometimes inclusions of tiny crystals of rutile, actinolite, and other minerals, visible to the naked eye.

System, the hexagonal system and hexagonal trapezohedral symmetry. The a-quartz modification, more stable at temperatures below 573° (at atmospheric pressure), crystallises in the trigonal system and trigonal trapezohedral *symmetry.* Crystal structures are rather simple. In each SiO_4 tetrahedron two oxygen ions are located somewhat above, and the other two somewhat beneath the silicon ion. The groups of tetrahedra lie in three layers at different heights; their relative positions above the plane of drawing are indicated by lines of different thickness and by figures alongside the circles. The tetrahedra are arranged in spirals, all coiling in the same direction.

The low-temperature a-quartz modification differs but little from quartz in crystal structure. It will be seen that in the polymorphous transformation of high-temperature into low-temperature quartz the centres of the silicon-oxygen tetrahedra are somewhat displaced, the lattice becoming more compact and its symmetry lower:

> The hexagonal axes become trigonal. The type of bonding between the tetrahedral groups

remains the same. Nor does the direction of spiral coiling change (to the left or to the right) in the course of transformation.

Habit: Crystals of /3-quartz, or rather the paramorphs of a-quartz after /3-quartz occurring in silica-rich effusives (liparites, quartz-porphyrys, etc.) as early magmatic segregations, have the hexagonal dipyramids habit with prismatic faces being greatly shortened or altogether absent. Though usually very small, in some rocks crystals 1 or even 2 cm across are found. Well-developed crystals of a-quartz occur only in cavities or in loose formations. Individual very large crystals weighing up to 40 tons are known. Although the crystal forms are extremely diverse, they have characteristic prismatic faces *m* (1010), often horizontally striated; rhombohedral faces *r*(1011) and 2(0111). Some individuals have equally developed faces of principal rhombohedra, the crystals assuming then the shape of a "hexagonal dipyramid". When etched, the faces of one rhombohedron remain shiny, those of the other become dull. The position of the faces of a trigonal trapezohedron {5161}, and dipyramid {1121} distinguishes the right-handed and the left-handed quartz. These faces are confined to prism edges (120°) and occur at the top and bottom on the opposite sides of the edge.

Besides the phanerocrystalline modifications of a- and quartz, there are cryptocrystalline varieties of fibrous texture: *chalcedony* and *quartzine* differing between themselves only in optical properties (e.g., in optic sign).

Twins, according to different laws, are very frequent.

1. *Dauphine* twins are so closely intergrown that they look like crystals, with the difference that they have double the number of the #-trapezohedron faces, that develop from each other by rotation through 60° instead of 120° about the vertical axis. The prism faces of both individuals merge together, and the faces of the r-rhombohedron coincide with those of the

z-rhombohedron. The optical axes of both individuals are parallel and, therefore, the twinning can be revealed only by etching. The composition faces are sinuous.

Brazilian twins differ from Dauphine twins in that they have double the number of trapezohedral faces, which are developed in a different way, namely by reflection in a vertical plane. The twinning structure is revealed by etching. Unlike the Dauphine structure, the composition faces are straight. Brazilian twins may also be identified optically.

Japanese twins are arranged parallel to the trigonal dipyramid {1122}; the individuals are inclined to each other at an angle of 84°34'.

Aggregates: Druses of quartz crystals are very common in cavities where they are sometimes intergrown with crystals of other minerals. Massive quartz is composed of granular aggregates. The structure of compact aggregates is readily recognised in thin polished sections under the microscope with crossed nicols. Chalcedony whose structure is cryptocristalline (fibrous) often occurs as crusts, reniform masses or spherolites, but mostly in nodules known as flints. Agates (siliceous geodes) have a concentric-banded structure due to alternating of variously coloured layers of chalcedony and sometimes quartz. The centres of the geodes are often composed of crystalline granular quartz, sometimes in the form of crystal druses.

Colour, Quartz occurs in a wide variety of colours, but most widespread are colourless, milk-white, and grey varieties. Transparent and translucent beautifully coloured varieties have names of their own: (1) *rock crystal,* colourless, water-transparent crystals; (2) *amethyst,* violet varieties; (3) *rauchtopaz,* a smoky transparent quartz tinted grey or brown; (4) *morion,* a black crystalline variety; (5) *citrine,* golden-yellow or lemon-yellow crystals. In addition to the transparent varieties, there exist definitely allochromatic quartz crystals whose colour is due to inclusions of foreign minerals; these also have individual names: *prase,* a greenish variety with inclusions of green acicular actinolite; *aventurine*, a yellowish of brownish-red quartz with a glimmering chatoyancy caused by tiny glistening scales of mica, specularite (Fe_2O_3), etc.

Milk-white quartz masses widespread in hydrothermal deposits sometimes owe their colour to abundant microscopic occlusions of liquids and gases. The liquid occlusions almost always contain a mobile gas bubble by which one easily recognises the presence of the liquid. Sometimes they even contain NaCl crystals. On heating to a certain temperature the microcrystals dissolve and the bubbles disappear. On cooling the different phases reappear. In this way it is possible to determine approximately the temperature at which the liquids and gases were trapped in the process of quartz crystallisation. Quartz may also be milk-white due to intense fracturing caused by dynamic factors (just as transparent ice turns milk-white when struck with a hammer).

Chalcedony, more often than crystalline quartz, has different colours and tints: milk-grey, bluish-black (sap-phirine), yellow, red, orange (carnelian), brown (sard), green (plasma), apple-green due to the presence of nickel compounds (chrysoprase), and green containing scattered red spots (heliotrope), etc. *Agates* or *onyxes* often composed of the thin, variously coloured, concentric or parallel bands of chalcedony sometimes show most diverse combinations of tints: black with white, brown with white (sardonyx), red with white (carneolonyx), etc.

Lustre, vitreous; for chalcedony, waxy to dull. Optical constants. Uniaxial, positive n_x= 1.553 and n_y= 1.544. Capable of rotating the polarisation plane, the direction of rotation depending on the type of quartz, i.e., right-handed, or left-handed.

Hardness, 7. *Cleavage*, either absent or imperfect rhombohedral Fracture, conchoidal. Specific gravity, 2.5 to 2.8, for pure varieties, with less close packing, has a somewhat lower specific gravity. Other properties. Transmits ultraviolet rays. Piezoelectrical, i. e., develops electric charges when subjected to pressure. The two-fold axes are the electrical ones, the end of each twofold axis terminating at the edge truncated by trapezohedral faces, developing a negative charge, and the opposite a positive charge. Melting point 1713°. A product of solidification of molten quartz is quartz glass (amorphous quartz) which has a number of special properties such as acid-resistance, low expansion coefficient, transparency to ultraviolet rays, etc.

Diagnostic Features: Quartz crystals are readily identifiable by their characteristic form. In compact aggregates quartz is recognised by high hardness, conchoidal fracture, and absence of cleavage.

Chalcedony differs from similar minerals (opal, smithsonite $ZnCO_3$, cryptocrystalline fluorite, etc.) mostly by hardness. Is also readily distinguishable by optical properties.

Infusible in the blowpipe flame. Insoluble in acids except HF with which it forms SiF_4, a volatile compound. Soluble by alkalis; chalcedony is strongly attacked by KOH.

Genesis and Occurrence. Being widespread in nature, quartz enters into the composition of rocks and ore deposits of most diverse genesis.

1. It is an essential constituent of many intrusive and effusive acid *igneous* rocks, along with feldspars and micas (granites, gneisses, quartz rocks, etc.). Porphyry quartz crystals in acid effusive rocks have a crystalline-banded texture and often contain inclusions of volcanic glass.
2. Large crystals (rauchtopaz, morion, amethyst, etc.) occur in cavities in *pegmatites* and also in association with feldspars, muscovite, topaz, beryl, tourmaline, and other minerals. Intergrowths are often observed with large individuals of potassium feldspars—orthoclase and microcline, whose polished specimens display patterns resembling Hebrew characters. It is characteristic that all quartz inclusions in each feldspar crystal have a uniform optical orientation.

3. Being a constant mineral of veins, in which it occurs in large masses, quartz is common in *hydrothermal* deposits extremely numerous throughout the world, in which it occurs in association with most diverse minerals: cassiterite, wolframite, gold, molybdenite, pyrite, chalcopyrite, tourmaline, calcite, chlorites, etc. Quartz practically always contains microinclusions of gases, liquids, and mineral solids.

Agates and onyxes as amygdules (secretions) of most diverse forms and sizes are widespread in many effusive rocks—melaphyres, bazalts, andesites, etc. (altered blister lavas). The formation of these varieties was due to the migration of concentrated colloidal silica solutions into the above rocks at the latest stage of hydrothermal processes.

4. In the course of *exogenous* processes, quartz and chalcedony, in the form of fine-granular aggregates, derive from the dehydration and crystallisation of silica gels. Sometimes, quartz crystals are formed in cavities, e.g., in the fissures of exogenously silicified limestones, serpentinites, and other rocks.

The formation of chalcedony in these conditions is much more extensive since it is due not only to the dehydration of silica gel, but also takes place independently. It has been deposited, for example, after opal in the voids between colloform segregations of the latter. Chalcedony occurs most widely as flints in limestones, and as nodules in the crust of weathering, resembling the amygdules and geodes of homogeneous chalcedony of hydrothermal origin found in effusive igneous rocks.

5. Large masses of quartz are formed by dehydration of opal-bearing sedimentary rocks which gives rise to the so-called jaspers and laminated hornfels (very close-grained quartz and quartz-chalcedony rocks).

Most interesting, however, both from the mineralogical and the practical standpoints are veins of the "Alpine-cleft" type in lenticular fissures developing in the course of metamorphic and schist-forming processes. In large cavities and fissures, the walls may be lined with well-formed, often large, crystals of quartz accompanied by chlorites, feldspar, rutile, brookite, etc. These veins and crystals have many specific features. The rock crystal individuals have most diverse orientation in respect of the walls of the cavities. It has been found that their orientation depends on that of the quartz grains exposed in the wall rock which served as the initiators for the large crystals that have grown in the cavity. Undeformed rock crystals are piezoelectric.

Being resistant to weathering and chemically inert, quartz accumulates as detrital grains in placers and sedimentary rocks (sandstones, quartzites).

Quartz pseudomorphs have been found after most diverse minerals: calcite, barite, gypsum, feldspars, olivine, etc., and also after animal and vegetal remains.

Since quartz deposits are very numerous, we shall limit ourselves to the most important localities.

The following deposits in the USSR are highly interesting mineralogically: old pegmatite mines at *Murzinka, Lipovka,* and *Shaytansk* (northeast of Sverdlovsk) where druses of excellent rock crystals, smoky quartz, and amethyst occur associated with feldspars, tourmaline, iepidolite, etc. Pegmatite dikes containing morion and amethyst are widespread in the *Adun-Chilon* ridge (Trans-Baikal country), *Volhynia* (the Ukraine), etc. Quartz veins with rock crystal are also found along the *Aldan* River and in the *Pamirs.*

Sealing-wax-red banded jaspers used as ornamental stones are widespread in the Magnitogorsk area (the Urals). Well-known is the uniformly greenish-grey kalgan jasper, and also the jaspers occurring in the Orsk district (the Southern Urals), which display a wide variety of patterns.

Deposits of the so-called industrial agate occur at *Akhaltsykh* and elsewhere in Transcaucasia. They are genetically associated with sheets of effusive rocks.

Notable foreign localities for noble quartz varieties are (Minas Geraes *Brazil)* and *Uruguay* (mostly amethyst), *Madagascar* (rock crystal), the *Swiss Alps*, etc. All these deposits are in the main associated with pegmatite dikes and quartz veins of the Alpine-cleft type. The best agates are found in India, Brazil, Uruguay, Germany (Ober-stein, Rhineland, etc.).

Uses: Quartz and chalcedony have many applications.

1. Transparent and beautifully-coloured varieties are used as gemstones.
2. Colourless rock crystal is used in the manufacture of optical instruments.
3. In precision mechanics some varieties of these minerals (especially industrial agate) are used as fulcrums and step bearings, watch jewels, etc.
4. In electronics, for the manufacture of piezoelectric crystal plates for frequency stabilisation, piezoelectric resonators, etc. Only untwinned and perfectly homogeneous crystals (rock crystal, smoky quartz, morion) with a minimum distance of 3.5 cm between the opposite prismatic faces, may be used for these purposes.
5. Quartz is used in making highly refractory and acid-resistant laboratory utensils, and in the manufacture of quartz lamps used in medicine as a source of ultraviolet radiation, since quartz is one of the few minerals transparent to ultraviolet rays.
6. In glass and porcelain making use is made of pure quartz sands with a maximum iron content of 0.002 per cent.
7. Another use of quartz is in the preparation of silicon carbide or carborundum, (SiC), a high-grade abrasive, which is harder than corundum.
8. Fine-grained quartz sands are used in sandblasters for polishing metal and stone articles, for cutting certain rocks, and for other purposes.

9. Sandstones composed of cemented rounded quartz grains, as well as their metamorphosed varieties (quartzite rocks) are used as a building material.

Trydimite, SiO_2

Trydimos is the Greek for triple; so named because of its frequent occurrence as trillings. The high-temperature modification, which crystallises in the hexagonal system, in the unstable range readily changes to low-temperature a-trydimite which, though similar in crystal structure, crystallises in the ortho-rhombic system. Under atmospheric pressure a-trydimite with time may gradually change into the more stable a-quartz.

The crystals of a-trydimite, which is more stable at low temperatures, occur as pseudohexagonal plates or more often as trillings with an angle of 35°18' between the plates. There occur also imbricate and rosette-like aggregates. The colour of a-trydimite is white, greyish-white; sometimes colourless. Lustre, vitreous $n_x = 1.473$, $n_y = 1.470$, and $n_z = 1.469$. Hardness, 6 to 7. Cleavage, imperfect. Specific gravity, 2.30.

Occurs mostly in the cavities of acid effusive rocks: trachytes at Siegengebirge (Germany), andesites at San Cristobal (Mexico), volcanic products of Mt. Vesuvius, etc.

Cristobalite, SiO_2

Named after the place of occurrence San Cristobal (Mexico). Polymorphous modifications: high-temperature cristobalite in the cubic system and low-temperature a-cristobalite. The latter crystallises in the tetragonal system, but is at the same time pseudocubic.

Crystals of cristobalite are octahedral in habit, more seldom cubic or skeletal. Twins on (111) are known. Colour, milk-white. Lustre, vitreous. $n_y = 1.49$. Hardness, 7. Specific gravity, 2.27.

The mineral is readily obtained during the making of quartz glass, and also crystallises during the firing of silica brick. Is found in rapidly-cooled effusive rocks, often together with a-trydimite: in the andesites of San Cristobal (Mexico), the lavas of Main in Rhineland (Germany), Yellowstone Park (U.S.A.) — as globules (spherolites) up to 1 mm in diameter in obsidian (volcanic glass), and also in cavities sometimes with trydimite plates growing over small cristpbalite crystals.

Other ways of formation of cristobalite are known as well, e.g., through the action of basaltic magma on quartz-bearing sedimentary rocks (sandstones). Cristobalite is in this case formed out of quartz at high temperatures. Such formations have been recorded from upper Tertiary clayey sandstones in Western Georgia (U.S.S.R.).

Opal, SiO_2-nH_2O

Typical solid hydrogel. The origin of the name is unknown. Varieties differing in physical properties are (1) *precious opal,* opalescent; *hydrophane,* light and very porous, clouded when dry and transparent in water; (3) *hyalite,* stalactitic or globular.

Chemical Composition: Chemical composition extremely variable. Water content from 1 to 5 per cent, rarely up to 34 per cent. All opals lose some water in a desiccator. This process is especially rapid in the first few days. On heating some opals lose most of their water at temperatures below 100°, and others at higher temperatures (between 100 and 250°). The causes of this difference in the strength of the water bond is as yet obscure.

Usually occurs in compact glass the principal constituent of organisms making up shells of diy of sinter-like siliceous spicules of sponges, the opal skeletons of radiolaria, partly of some foraminifera and bryozoan feeding on colloidal silica solutions (sols). Owing to the presence of silica, the skeletons of these organisms are well preserved as fossils even in most ancient sediments. Silica hydroxides also enter into the composition of the stalks of cereals, hard nodules of equisetaceous plants, especially of bamboo and other plants that absorb silica sols from the soil through their roots. Silica sols are capable of impregnating dead trunks of trees and completely replacing the wood with opal so that the woody structure is faithfully retained (wood opal).

Colour Pure opal is colourless, but owing to impurities, especially iron and other chromophores, it occurs in a variety of tinges: yellow, brown, red, green, and black. *Lustre,* vitreous; for porous varieties waxy and earthy. Translucent varieties opalescent (the term itself derives from the name of the mineral). *Hardness,* 5 to 5.5 (for varieties with low water content up to 6). Brittle. *Specific gravity,* 1.9 to 2.5 (depending on the content of water and absorbed heavy matter).

Diagnostic Features: Compact opals are characterised by glassy appearance and sinter-like shapes. Opal strongly resembles chalcedony externally, differing from it by lower hardness and water content.

Infusible in the blowpipe flame, but often decrepitates violently. Yields water in the closed tube. Insoluble in acids. Rather easily soluble in KOH and HF. Dehydrated opal dissolves in soda with effervescence (due to evolution of CO_2).

Genesis and Occurrence: Opal is frequently deposited from *hydro-thermal* springs and geysers in volcanic regions as sinters (siliceous tuff, geyserite) and of white translucent botryoidal masses with a pearly lustre. Also occurs in cavities and fissures in effusive rocks, sometimes as amygdules and geodes.

The bulk of opal is formed, however, through exogenous decomposition of silicates as a result of the *weathering* of rocks of most diverse composition, but most frequently ultrabasic. The silica liberated in the disintegration of the crystal structures of silicates first changes into a sol, upon the coagulation of which it is precipitated as sinter-like nodules in the eluvial zone, or is deposited metasomatically, often together with hydroxides of iron, aluminium, and other elements on various bedrocks.

Large opal beds are formed by *sedimentation* through the coagulation of silica sols brought by river waters to near-shore zones of marine basins. These formations include the so-called gaizes, tripolis, diatom-ites, kieselgur (mountain meal), etc., that are loose or finely porous, sometimes more or less hard rocks, often altered at the surface into chalk-like masses by the mechanical action of freezing water they are saturated with.

Finally, mention should be made of the fossilising role of silica colloids in the metasomatosis of plant remains such as the trunks of trees. Whatever its mode of formation, opal gradually changes into chalcedony or quartz.

Sedimentary opal-bearing rocks (gaizes, tripolis, and diatomites) are especially widespread in Tertiary shallow-water deposits in Ulyanovsk, Saratov and other regions of the European part of the USSR, along the eastern slope of the Urals; in the Trans-Caspian and the Trans-Caucasian regions and elsewhere.

Outside the Soviet Union, the Sedimentary opal-bearing rocks (gaizes, tripolis, and diatomites) are especially widespread in Tertiary shallow-water deposits in Ulyanovsk, Saratov and other regions of the European part of the USSR, along the eastern slope of the Urals; in the Trans-Caspian and the Trans-Caucasian regions and elsewhere. Outside the Soviet Union, the most important localities are Bohemia, Italy, and Tripoli (hence the name of tripoli). Deposits of precious opals are known on the southern slope of the Carpathians, at Kosice (in cavities in andesite), in New South Wales and Queensland (Australia) in sandstones occupying an enormous territory, and elsewhere.

Uses: Precious opals are ornamental stones. Tripoli is used for polishing metals and stones, and other purposes; kieselgur for the manufacture of filters, ceramics, light-weight bricks, etc.

Summary: Thus all the natural modifications, except opal and lechatelierite (amorphous quartz glass), are characterised by crystal structures comprised of silicon-oxygen tetrahedral networks which share corners, i.e., each silicon ion is surrounded by four oxygen ions, and each oxygen ion by two silicon ions. The structures differ in the spatial orientation of the tetrahedra.

Hydroxides of Hydroxyl-containing Oxides

The most important minerals in this class are the so-called hydrates or hydroxides, i.e., compounds of metals with the hydroxyl OH radical, completely or partly replacing the oxygen ions in oxides. Thus, $Mg(OH)_2$ corresponds to magnesium oxide —MgO; 2AlO(OH) or $2Al(OH)_3$ correspond to aluminium oxide — Al_2O_3, etc.

Most of these compounds crystallise in layered structures with the hexagonal or hexagonal-type closest packing of $[OH]^{1-}$ ions. This feature explains many common physical properties of these minerals. The optimal cations of the compounds with layered structure are Mg^{2+}, Fe^{2-}, Mn^{2+}, Al^{3+}, Fe^{3+} and Cr^{3+}.

Sedimentary opal-bearing rocks (gaizes, tripolis, and diatomites) are especially widespread in Tertiary shallow-water deposits in Ulyanovsk, Saratov and other regions of the European part of the USSR, along the eastern slope of the Urals; in the Trans-Caspian and the Trans-Caucasian regions and elsewhere.

Brucite Group

The group includes hydrates of divalent metals, such as Mg, Fe, and Mn crystallising in the trigonal or hexagonal system. Only brucite will be discussed here.

Brucite, $Mg[OH]_2$

Chemical Composition: MgO 69 per cent, H_2O 31 per cent. May contain isomorphous admixtures of Fe (ferrobrucite) and Mn (manganobrucite).

System, trigonal; *symmetry* ditrigonal scalenohedral. Crystal structure, typically laminated. The anion of this structure is a dipole hydroxyl group $[OH]^{1-}$. In the crystal structure these groups follow the pattern of the closest hexagonal packing. Each layer consists of two flat sheets of hydroxyl ions parallel to the (0001) plane. Sandwiched between these sheets is a sheet of Mg^2 cations that occupy all the *octahedral* sites between the two OH sheets, i.e., each Mg ion is surrounded by six hydroxyl anions, viz., three $[OH]^{1-}$ ions of one sheet and three [OH] ions of the other sheet. The perfect cleavage is parallel to the planes between these treble layers which are linked by weak residual forces.

Habit: Thick-tabular. Commonly as foliated talc-like aggregates. Sinter-like metacolloidal formations also occur sometimes.

Colour, white, rarely greenish or colourless. *Lustre,* vitreous on fracture and pearly on cleavage surfaces. n_x = 1.580 and n_y= 1.559. *Hardness,* 2.5. *Cleavage,* {0001} excellent. Thin, easily separated foliae flexible. *Specific gravity,* 2.3 to 2.4.

Diagnostic Features: Brucite is similar to talc —$Mg_3(OH)_2Si_4O_{10}$, pyrophyllite — $Al_2(OH)_2Si_4O_{10}$, and hydrargillite $Al(OH)_3$. Differs from them sharply by ready solubility in hydrochloric acid.

Infusible, but glows, before the blowpipe flame. Dissolves readily in acids without effervescence. Gives off water in the closed tube.

Genesis: Is formed through hydrolysis of dissolved magnesium compounds in a strongly alkaline environment. Often occurs as thin vein-lets and selvages in fissures in serpentinites, these being the product of hydrothermal alteration of magnesia-rich ultrabasic igneous rocks (dunites and peridotites). Associates are serpentine, magnesium hydro-carbonates, aragonite, etc. Sometimes found in dolomitised limestones with calcite, hydromagnesite, and periclase, often as pseudomorphs after periclase, probably of hydrothermal origin. Has also been recorded from strongly alkaline soils.

In a more acidic environment at the surface, brucite readily carbonatises.

Uses: When found in commercial quantity, brucite is a source for metallic magnesium.

In the USSR, occurs in all unweathered serpentinite masses in the Urals, the Caucasus, and in Siberia. A fibrous asbestos-like variety, nemalite, occurs at the Bazhenovskoye asbestos deposit (town of Asbest, east of Sverdlovsk).

Hydrargillite Group

This group comprises the so-called trihydrates of certain trivalent metals. The most widespread in nature is $Al(OH)_3$. Ferric hydroxide, $Fe(OH)_3$, which forms first in the hydrolysis of iron salts, is unstable and rapidly dehydrates. Fe^{3+} ions in small quantities may be admixed to $Al(OH)_3$ only isomorphously. Aluminium hydrates also contain Ga^{3+} ions as an isomorphous admixture in quantities much higher than its clark indicates.

Hydrargillite, $Al(OH)_3$

The name derives from the Greek *hydor* — water and *argillos* — white clay. First established in the last century in the Urals. Synonym: gibbsite.

Chemical Composition: Al_2O_3 65.4 per cent, H_2O 34.6 per cent. May contain isomorphous admixtures of Fe_2O_3 (up to 2 per cent) and Ga_2O_3 (up to 0.006 per cent). Chemical analyses often show almost complete correspondence of its composition to the formula.

System, monoclinic; *symmetry* prismatic. Crystal structure layered, close to the distorted structure of brucite. Its specific features are explained by the fact that the aluminium ion has a greater charge and smaller radius than the magnesium ion. In hydrargillite, as in brucite, each layer consists of two closely packed sheets of hydroxyl ions with a sheet of Al^{3+} cations. Since the aluminium ion carries a greater charge than the magnesium ion, less Al^{3+} cations will be required to cancel out the negative charge of the hydroxyl ions. Therefore AP^{+} cations do not occupy all the octahedral sites between the OH sheets, as in the structure of brucite, but only two-thirds of them. As a result, the Al^{3+} ions are arranged not as centred hexagons, as in the brucite structure, but as simple hexagons. If such a hexagon is outlined, it will be easily seen that in the brucite lattice it would include six hydroxyl ions and three complete Mg^{2+} ions (one central ion and six peripheral ones, each having 2/6 of the valence), i.e., the hexagon will be expressed by the formula $Mg_3(OH)_6$. On the other hand, in the hydrargillite structure, which has the same number of hydroxyl ions, the formula of the hexagon will have only two Al^{3+} ions, i.e., will be $Al_2(OH)_6$. Another difference lies in the manner of linkage of the treble layers: while in the brucite structure each $(OH)^{1-}$ ion is located against the centre between the three $(OH)^{1-}$ ions of the adjacent layer, in the hydrargillite structure each $(OH)^{1-}$ ion of one layer lies opposite to each $(OH)^{1-}$ ion of the next layer.

Habit: Hexagonal-tabular. Occurs in complex twins according to several laws, usually along (100) and (110). More often as radial-foliated aggregates, sometimes as sinters or pisolitic and globular concretions. The bulk of hydrargillite at the earth's surface occurs as thin-scaly or cryptocrystalline masses.

Colour, white or slightly greyish, greenish, and reddish. *Lustre,* vitreous, pearly on *cleavage* surfaces. $n_x = 1.587$, $n_y = 1.566$, and $n_z = 1.566$. Hydrar gymte cry *Hardness,* 2.5 to 3.5. *Cleavage,* parallel to the basal pinacoid {001} excellent. *Specific gravity,* 2.43. Other properties. On heating changes to boehmite, AlOOH, and on ignition up to 950° to y-Al_2O_3 (with a spinel-type cubic structure).

Diagnostic Features: Characteristic features of hydrargillite are excellent cleavage, vitreous lustre, and low specific gravity. Differs from similar diaspore by hardness (6 to 7 for diaspore), from mica by specific gravity and optical properties (micas are optically negative).

Infusible in the blowpipe flame, but yields OH, and becomes white and opaque. The hydroxyl is lost in two stages: (1) at a temperature of 202-196° (conversion to boehmite) and (2) at a temperature of 510° (further dehydration). Ignited hydrargillite becomes bright-blue when moistened with Co solution. Like hydrates of other metals, dissolves in dilute hydrochloric acid.

Genesis and Occurrence: Derives from the decomposition and hydrolysis of aluminium-bearing silicates, partly in the course of *hydro-thermal* processes (at a comparatively low temperature), but chiefly from *surface weathering*, mainly under tropical and subtropical climatic conditions.

Hydrargillite of hydrothermal origin occurs rather rarely and in negligible amounts. It has been observed in certain endogenous deposits as one of the last minerals to crystallise from low-temperature hydrothermal solutions.

In the process of weathering in tropical countries aluminium hydroxides, including hydrargillite, are usually formed together with iron hydroxides. Hydrargillite is present in far greater amounts in the so-called *laterites*, i.e., products of weathering widespread in modern tropical countries as a mantle over bedrock, and comprised mainly of hydroxides containing Fe_2O_3 and, to a smaller extent, Al_2O_3 and SiO_2. It is also found in *bauxites* which are essentially aluminium hydroxides and are also formed by *sedimentation* in bodies of water, apparently as a result of weathering and denudation of aluminium-bearing rocks.

The formation of bauxites in certain deposits is also attributed to decomposition of limestones and dolomites under exogenous conditions, the assumption being that the clayey residues were further decomposed in alkaline environment and the silica thus liberated carried away.

In the course of regional metamorphism hydrargillite loses its water changing to diaspore, and to corundum (emery) at greater depths, or in the presence of SiO_2 to alumosilicates.

In the Soviet Union, hydrargillite occurs as an accessory of diaspore in bauxite deposits in the *Tikhvin* district, Leningrad Region, and elsewhere. Hydrargillite crystals up to 5 cm across, and hence of mineralogical value, have been recorded from hydrothermal formations of the *Shishim* and *Nazyam* mountains in the Zlatoust district (Southern Urals) in metamorphic schists, and also as the alteration product of nepheline in the pegmatites of Vishneviye mountains (Middle Urals).

Uses: Hydrargillite, like such constituents of bauxite as diaspore and boehmite, is a source of alumina, for making aluminium. Bauxites used for this purpose should not contain more than 10 to 15 per cent of silica.

The chemical industry uses slightly ferruginous (light-coloured) bauxites to obtain aluminium salts, mostly sulphates. Bauxites of lower grades are used for the manufacture of abrasives (alundum and aloxite) and refractories for high-temperature furnaces (with a melting point above 2000°). Increasing use is made of special grades of bauxites as adsorbents in refining of mineral and vegetable oils, in the manufacture of paints, as catalysts, etc.

Sassolite, $B(OH)_3$

System, triclinic; *symmetry,* pinacoidal. Occurs in fine-scaly, hexagonal-tabular colourless or slightly coloured crystals. Lustre, vitreous; pearly on cleavage surfaces. Optically negative. $n_x = 1.459$, $n_y = 1.456$, and $n_z = 1.340$. *Hardness,* 1. *Cleavage* {001}, excellent. *Specific gravity,* 1.48. Readily fusible. Colours the flame green. Poorly soluble in cold water.

Occurs as white efflorescences —products of fumarole activity in volcanic areas: in large amounts on Volcano Island (the Lipari Archipelago, northeast of Sicily), in cracks of Avacha Volcano on Kamchatka, and elsewhere. Was first found at Sasso in Tuscany (Italy). Dissolved boron hydroxide is often recognised in oil reservoir waters and in the products of mud volcanoes (on the Kerch Peninsula and along the Caucasian coast of the Caspian Sea).

LEPIDOCROCITE-GOETHITE GROUP

Included in this group are the so-called monohydrates of trivalent metals: Al, Fe, Mn, and Co. The compounds are dimorphous and crystallise in the orthorhombic system.

The lepidocrocite sub-group (FeOOH) includes its aluminium analogue — boehmite (AlOOH), while the goethite ($HFeO_2$) sub-group includes diaspore ($HAlO_2$). Manganite the empiric formula of which is similar to the above compounds will also be described here, although otherwise it has little in common with them.

Boehmite, AlOCH.

Named after Boehm who established the presence of this mineral in bauxite by X-ray studies; the mineral resembles lepidocrocite (FeOOH) in crystal structure.

Chemical Composition: The same as of diaspore. Al_2O_3 84.97 per cent (according to formula). Contains also SiO_2, Fe_2O_3 (probably as a mechanical admixture of opal and iron hydroxides), and also Ca_2O_3.

System, orthorhombic; *symmetry* rhombic dipyramidal. Crystal structure, analogous to that of lepidocrocite.

Habit: Occurs in cracks and pores in bauxite and also in the decomposition products of nephelme as crystal from Vishvery small lamellar or lenticular crystals 199) neviye mountains, usually in combinations: (010), (111), (113), {110}, etc. Characteristically, the faces {010} are lustrous, whereas the faces of other forms are dull. Usually occurs as cryptocrystalline masses or as colloform aggregates (in bauxites).

Colour, Colourless or white with a yellowish tinge. The average refractive index for cryptocrystalline varieties is 1.640-1.645. *Hardness,* 3.5. *Cleavage,* {010} perfect. *Specific gravity,* 3.01 to 3.06. On ignition changes to y-Al_2O_3 (cubic modification crystallising in the spinel structure).

Diagnostic Features: In view of the extremely small size of crystals, a debyegram is the most reliable means of identification. Differs from diaspore by lower refractive index and lower hardness.

Before the blowpipe does not fuse, but turns white, and splits along cleavage planes. Gives off water in the glass tube. Insoluble in acids.

Genesis and Occurrence: Until recently boehmite has been found only in *exogenous* bauxite deposits (cf. hydrargillite).

Lately it was discovered in the ancient crust of weathering of the Yakovlevskoye deposit (Kursk Magnetic Anomaly) in association with kaolinite overlying metamorphic schists of ferrous-magneso-aluminous composition. Still more recently, it was found as a low-temperature hydrothermal mineral in small crystals in cavities in pegmatites of alkaline rocks of the *Vishneviye mountains* (Middle Urals) associated with water-transparent hydrargillite on acicular zeolite (natrolite) crystals. It has probably formed through hydrothermal alteration of nepheline.

Lepidocrocite, FeO[OH]

Lepidos is the Greek for plate, *crocos*, saffron. Synonym: rubinglimmer.

Chemical Composition: Similar to that of goethite. According to the chemical analyses data, contains far less impurities than goethite. As the latter has varieties containing adsorbed water (in finely crystalline colloform masses) called *hydrolepidocrocite*, $FeO[OH]\text{-}3H_2O$.

System, orthorhombic; symmetry, rhombic dipyramidal. Crystal structure. As shown in 200, the Fe^{3+} cations are surrounded by six oxygen ions (the front and the rear anions have been left out) which occupy the apexes of distorted octahedra arranged in chains parallel to the *a* axis. The chains are laterally (through octahedral edges) linked with each other into continuous double sheets parallel to (010). It is bonded with two H cations, its valence being thus fully saturated. As for the O^{2-} anions within the double sheets, their valences are also fully saturated by the shares of valences supplied by Fe^{3+} cations (each O anion is surrounded by four Fe cations). Thus, two types of oxygen ions take part in the structure of lepidocrocite, and, therefore, the formula of this mineral must be written as FeO[OH]. Hydroxyl ions as such are absent. In the goethite ($HFeO_2$) structure, however, all the oxygen ions are chemically equivalent. The less close packing of ions in the lepidocrocite structure is reflected in its lower specific gravity.

Habit: Occurs in fine crystals, platy along (010), in cavities, but mostly in fine-scaly or fibrous aggregates. Sometimes in reniform aggregates on the walls of geodes.

Colour, dark-red to reddish-black, sometimes shows a golden tinge (in scaly masses). *Streak,* orange or brick-red. *Lustre,* adamantine. Thin polished sections transparent. n_x= 2.51, n_y= 2.20, and n_z= 1.94. *Hardness,* 4. *Cleavage,* {010} excellent and {001} perfect. *Specific gravity,* 4.09 to 4.10.

Diagnostic Features: Often mistaken for haematite, especially in massive aggregates. Separate laminae are readily identifiable by crystal form, red colour in transmitted light, orange or brick-red streak, and low specific gravity (as different from haematite). Infusible in the blowpipe flame; at high temperatures becomes black and magnetic. Gives off water in the closed tube. Soluble in hydrochloric acid.

Genesis and Occurrence: Lepidocrocite in platy crystals (ruby mica) occurs occasionally as one of the last-formed minerals in *hydrothermal* deposits, e.g., at *Siegen,* Westphalia (Germany), and in iron-ore vein deposits (in the cavities of quartz stringers) in the Angara-Ilim region (U.S.S.R.) and elsewhere. Much more often, lepidocrocite occurs in concentric layers of scaly aggregates in brown ironstone sinters or on the walls of geodes of *surface* origin: at the Poletayevskoye deposits (southwest of Chelyabinsk), near Lipetsk, and elsewhere.

Microscopic and X-ray examination of the products formed on prolonged heating shows that lepidocrocite is capable of altering to goethite. The conditions under which these transformations occur and under which various modifications of iron monohydroxide are formed have not yet been fully established. Possibly, the acidity of the environment (pH) in which iron hydroxides are formed plays an important part in this process.

Diaspore, $HAlO_2$

Diaspore is the Greek for dispersal (some specimens, when heated, crack and disintegrate into small particles). First established in the last century in the Urals.

Chemical Composition: Al_2O_3 85 per cent, H_2O 15 per cent. Certain varieties contain as isomorphous admixtures Fe_2O_3 (up to 7 per cent), Mn_2O_3 (mangandiaspore), Cr_2O_3 (up to 5 per cent), and SiO_2 (up to 4 per cent). May also contain up to several hundredths of per cent of Ga_2O_3.

System, orthorhombic; *symmetry,* rhombic dipyramidal. Crystal structure characterised by the close hexagonal packing of O^{2-} ions, with the Al^{3+} ions occupying the octahedral voids, i.e., between six oxygen ions. Each oxygen ion is linked with three Al ions, i.e., the coordination numbers for Al and O are the same as in the crystal structure of rutile (6 and 3). The H^1 protons are probably located between the pairs of oxygen ions, and owing to their infinitesimal size, do not need any space of their own within the crystal structure.

Thus, the structure of diaspore is practically composed of Al and O ions in the 1:2 ratio (as in rutile). Its similarity to rutile, as demonstrated by N.V. Belov, is also structural with the difference that in diaspore (goethite), instead of single columns, there are pairs of closest-packed columns parallel to the *c* axis. Hence the platy habit, cleavage, and geniculate twins at an angle of about 122°.

Habit: Platy, sometimes tabular and frequently columnar parallel to the *c* axis. Faces striated vertically. Usually occurs in foliated or scaly aggregates.

Colour, yellowish-brown, white, light-violet, greenish-grey. Lustre, vitreous; pearly on cleavage surfaces. *Streak,* white. $n_x = 1.750$, $n_y = 1.722$, and $n_z = 1.702$. *Hardness,* 6 to 7. Very brittle. *Cleavage,* {010} distinct. *Specific gravity,* 3.3 to 3.5. On ignition changes to a-Al_2O_3 (corundum).

Diagnostic Features: Foliated aggregates, high hardness (as compared with hydrargillite, micas, etc.). May be mistaken only for certain varieties of chloritoid (alumosilicate of Fe and Al), a mineral sometimes associated with diaspore and corundum in metamorphic rocks from which it is almost indistinguishable by external features. In contrast to chloritoid, diaspore is not soluble in sulphuric acid, and has somewhat different optical properties.

Infusible in the blowpipe flame. Insoluble in acids and KOH. Becomes soluble in sulphuric acid only after strong ignition. On heating in a test tube disintegrates into small white scales.

Genesis and Occurrence: Occurs rarely in *contact-metasomatic* and *hydrothermal* deposits in marbled limestones in association with corundum, muscovite, haematite, rutile, etc.

Large masses and scaly aggregates occur in *exogenous* bauxite deposits in association with hydrargillite, boehmite, etc.

Often found in *metamorphic* rocks with corundum, chloritoid, and other minerals (in emery deposits, probably as a product of bauxite metamorphism), sometimes in crystalline schists as a rock-forming mineral, accompanied by disthene and other minerals.

In the USSR, occurs associated with corundum and alumosil-icates, in emery rock deposits in the Urals such as Kossoy Brod (Sverdlovsk district), in marbled limestones along the Borzovka River (Kysh-tym district), and also at Aktash (Uzbekistan) in association with du-mortierite, pyrophyllite, alunite, corundum, etc.

Large plates and crystals occur in the emery mines of Chester, Massachusetts (U.S.A.) and on the Island of Naxos (Greece).

Goethite, $HFeO_2$ and Limonite (hydrogoethite), $HFeO_2$-nH_2O

Named after Goethe (1749-1832). At first the mineral was called onegite, after the place of discovery (Lake Onega), but since its properties were not described, the name has not survived in mineralogical literature. Synonym: acicular iron ore (in German literature).

The name of limonite derives from the Greek *lemon* for meadow (allusion to meadow and bog ores).

Chemical Composition: Fe_2O_3 89.9 per cent, H_2O

10.1 per cent. The water content is often in excess of that given by the formula and may be as high as 12 to 14 per cent for limonite. For this reason several mineral species were formerly distinguished according to water content and certain physical properties. X-ray data show that there is actually only one compound with a ratio of $Fe_2O_3 : H_2O = 1:1$, possessing a definite crystal structure. All the iron hydroxides richer in water are actually hydrogels containing varying amounts of adsorbed water, depending on the degree of dehydration.

The so-called *turgite,* according to X-ray and thermal analyses, is a mixture of goethite with hydrohaematite and, therefore, not a mineral in its own right.

The accumulations of natural iron hydroxides are mostly mixtures of goethite with limonite, and also with silica hydroxides, argillaceous matter, etc. These mixtures are usually called *brown iron ores.*

System, orthorhombic; *symmetry,* rhombic dipyramidal. Crystal structure, analogous to that of diaspore.

Habit and Aggregates: Rare acicular or columnar crystals. Occasional twins similar to geniculate twins of rutile along (Oil). Usually in sinter-like, reniform, or stalactitic forms of thin radial- or parallel-fibrous texture on fracture (needle iron ore) or in porous, spongy, slagy or pulverulent masses. Pseudomorphs after crystals of pyrite and other iron sulphides are quite frequent. Also occurs in the forms of oolites, pisolites, concretions, and geodes.

Colour of limonite and goethite is dark-brown to black. Pulverulent or ochrous limonite which often derives from the physical weathering of compact black limonite and iron silicates is rather light yellow-brown in colour. From comparative chemical and X-ray evidence this ochrous variety does not differ materially from compact limonite. Streak of goethite is brown with a reddish tinge. That of limonite is mostly light-brown or yellow-brown. Lustre of goethite adamantine to sub-metallic. Goethite is often found on the surface of reniform or stalactitic limonites as thin lustrous pitch-black crusts. n_y = 2.35 to 2.39. *Hardness,* 4.5 to 5.5 for goethite and 4 to 1 for limonite, depending on physical condition. *Cleavage* of goethite {010} perfect. *Specific gravity,* 4.0 to 4.4 for goethite and 3.3 to 4.0 for limonite.

Diagnostic Features: Goethite and limonite are rather easily identifiable by colloform formations, brown streak and yellow-brown ochrous selvages.

Fusible in the blowpipe flame; upon prolonged heating become strongly magnetic. In the glass tube give off water, and turn red, altering to anhydrous Fe_2O_3. Slowly dissolve in hydrochloric acid.

Genesis and Occurrence: In small acicular or columnar crystals goethite very rarely occurs as an *endogenous mineral* in cavities of melaphyres, in amethyst geodes on Volk-Ostrov (an island on Lake Onega); sometimes in hydrothermal deposits as one of the most low-temperature minerals in cavities, e.g., at Pfibram (Czechoslovakia) in association with earlier sphalerite and pyrite untouched by weathering.

In the main, however, goethite and limonite are widespread as *exogenous* minerals and almost exclusively as colloform or metacolloidal masses. They are formed for the most part through the hydrolysis of salts.

Large masses of brown iron ores are found in the oxidised zones of sulphide deposits. These are the so-called *iron hats* (gossans) — loose, lumpy, or compact masses chiefly composed of limonite, goethite, sometimes lepidocrocite, etc.

Large masses of iron hydroxides accumulate in sedimentary deposits of young (Tertiary) brown iron ores accumulated in marine and lacustrine basins. The precipitation of iron hydroxides as well as of other hydroxides in the near-shore zones of these basins is evidently due to the coagulation of colloidal solutions brought by surface waters in marine waters through the action of electrolytes, and in freshwater lakes through the life activity of ferrobacteria. In addition to that, in certain sedimentary deposits iron hydroxides are formed as a result of recent oxidation of the carbonate and silicate iron ore facies in the oxidised zone.

Thus, limonite and goethite are formed almost exclusively at the very surface of the earth in the conditions of free access of oxygen and moisture.

In the course of regional metamorphism iron hydrates lose their water and are converted to anhydrous oxides (haematite and magnetite).

Deposits of brown iron ores of diverse origin are very widespread. The USSR has many huge deposits.

Uses: Brown iron ores, as well as haematites and magnetites, are a very important source of iron and steel. In the blast-furnace, the ores are completely dehydrated and converted into very finely-porous masses. Since the rate of ore reduction greatly depends on the specific surface of the mass, these ores are more economical than magnetite and haematite ores. Therefore, their permissible minimum iron content is 35-40 per cent (instead of 50 to 60 per cent for compact magnetite and haematite ores).

Manganite, Mn—Mn—$O_2(OH)_2$, or $MnO_2Mn(OH)_2$.

The presence of di- or tetravalent (not trivalent) manganese was established through the study of the magnetic anisotropy of the ions.

Chemical Composition: MnO 40.4 per cent, MnO_2 49.4 per cent, H_2O 10.2 per cent. Impurities: SiO_2 (up to a few per cent), Fe_2O_3 (up to 1 per cent), negligible quantities of Al_2O_3, CaO, etc. The H_2O content in cryptocrystalline and oolitic masses of manganite, as well as in goethite, may be in excess of the limnite crystal. Crystal structure materially differs from that of goethite and still awaits deciphering.

Habit: Prismatic, columnar parallel to the *c* axis. Faces {110} are deeply striated vertically.

Aggregates: In cavities of hydrothermal deposits manganite occurs often in druses of columnar crystals. In sedimentary deposits occurs widely as finely-crystalline aggregates, also as oolites, and rarely as sinters.

Colour, black. Streak, brown. *Lustre,* sub-metallic. $n_x = 2.53$, $n_y = 2.24$ and $n_z = 2.24$. *Hardness,* 3 to 4. Brittle. *Cleavage,* {010} perfect. *Specific gravity,* 4.2 to 4.33.

Diagnostic Features: Crystals readily identifiable by columnar habit, characteristic striation of prism faces, and also by brown streak. The distinguishing feature of colloform water-abundant variety is brown colour of both the mineral and the streak. Positive identification requires X-ray photography.

Infusible in the blowpipe flame. In the closed glass tube gives off abundant water. Borax and salt of phosphorus beads in the reducing flame are violet (presence of manganese). Dissolves in concentrated hydrochloric acid with evolution of chlorine.

Genesis and Occurrence: Manganite generally forms under conditions of oxygen deficiency. Occurs as druses of crystals in certain *hydro-thermal* manganese deposits, being one of the last minerals to be formed in paragenesis with barite and calcite.

Occurs in quantity in *sedimentary* deposits as oolites or massive occupying an intermediate place between the facies of psilomelane-pyrolusite ores (compounds of tetravalent manganese), and of carbonate ores containing divalent manganese located farther from the coastline.

Sometimes is found as globules of radially fibrous texture in clays. Unstable in the oxidised zone, changing readily to cryptocrystal-line, almost anhydrous MnO_2, i.e., all the manganese is oxidised to tetravalent manganese. Therefore, manganese hats comparatively rarely contain manganite.

Huge masses of manganite are found at *Chiatury* as brown oolitic ores. As black globular concretions it is widespread at the *Nikopol* deposit (the Ukraine). Crystals, sometimes large, have been recorded from the hydrothermal deposits of *Ilfeld,* Hartz; *Ilmenau,* Thuringia (Germany), and elsewhere.

Uses: Along with psilomelane pyrolusite-ores, manganite is an important material for ferromanganese and other ferrous alloys (spiegel and silicospiegel) used in steel making.

Psilomelane Group

This group comprises diverse manganese hydroxides of complex and variable composition, occurring as cryptocrys talline and collo-forrn masses. Earlier they were all classified with amorphous substances. Subsequently, however, on the basis of X-ray studies, many were found to be crystalline. In crystal structure they differ sharply not only from the oxides and hydroxides of manganese, reviewed heretofore, but also from one another. The differences, according to V.I. Mikheyev's incomplete X-ray studies, depend on the cations in their composition.

Chemical Composition: Chemical composition of many of these hydroxides is still obscure. They are complex hydrates, the simplest of which is the "mono-hydrate" of manganese dioxide, MnO_2-H_2O. This, in V.I. Vernadsky's opinion, corresponds to natural metamanganous acid, H_2MnO_3, which can be easily obtained in the laboratory in the form of a chocolate-brown gel.

More complex compounds include, besides the MnO_2 and MnO manganese oxides, the following oxides: K_2O, BaO, CaO, sometimes PbO, ZnO, CoO, NiO, CuO, occasionally Li_2O, WO_3, P_2O_5, As_2O_5, etc. There also have been recognised colloidal mixtures of hydroxides of manganese with those of SiO_2, Fe_2O_3, Al_2O_3, and with organic substances. These compounds have many different names depending on metal oxides entering into their composition. But in field practice these mineral species are known under the general name of "psilomelane".

The mineral species can be identified positively by means of debyegraphs and chemical analyses. Thermographs are unfortunately almost useless in the diagnostics of these minerals. The soft pulverulent brown or black varieties of psilomelane were formerly known as "wad". Today this term has lost its meaning, since detailed investigation of the varieties with the aid of X-rays and chemical analyses has shown that this form includes many minerals of the most diverse composition: mineral species of the psilomelane group, and also manganite, vernadite, etc.

Pselomelane, wMnO-MnO$_2$-rcH$_2$O

The precise formula has not been established as yet. *Psilos* is the Greek for bald *melas* — black. In the United States it has been suggested that the term be used for the variety in which barium predominates, called by other authors roma-nechite. The reason for this is that the first specimens which were given the name of psilomelane proved to be rich in barium. However, the most widespread variety of the psilomelane group is poor in barium.

Chemical Composition: Chemical composition is variable. The MnO : MnO_2 ratio varies greatly according to the degree of oxidation of the manganese protoxide. The MnO_2 content is usually 60 to 80 per cent, whereas that of MnO only 8 to 25 per cent, and H_2O from 4 to 6 per cent (most of it evaporates at temperatures over 110°). Small amounts of BaO (up to a few per cent) are often present, sometimes alkalis, CaO, CoO, MgO, ZnO, and also SiO_2, Fe_2O_3, Al_2O_3 as foreign admixtures. WO_3 (up to 1 per cent, occasionally 5 to 8 per cent) is sometimes present, in which case the variety is called *tungomelane*.

System, orthorhombic. Usually occurs in sinter masses with a concentric banded structure or as fine crystalline aggregates. Often as dendritic coatings.

Colour, black, sometimes brownish-black. *Streak*, usually black. *Lustre*, sub-metallic, for friable varieties dull. *Hardness*, 4 to 6 (for compact varieties), varies depending on water content and physical state. Brittle. *Specific gravity*, 4.4 to 4.7.

Diagnostic Features: As a member of the psilomelane group identifiable by sinter aggregates, black streak, and characteristic manganese reaction. Positive identification, however, is possible only by means of chemical analysis. Infusible in the blowpipe flame. Dissolves in hydrochloric acid with evolution of chlorine. In the closed tube gives off water and oxygen. Gives a violet bead with borax and salt of phosphorus in the reducing flame.

Genesis and Occurrence: As other minerals of the group, is formed mostly exogenously in the *oxidised zones* of manganese ore deposits and in deposits of *sedimentary* origin. Is also found as a minor mineral in the manganese ores of *hydrothermal* origin.

In the oxidised zone it is mostly a secondary mineral derived from braunite, hausmannite, sometimes from manganous silicates and carbonates (together with vernadite), but occasionally directly through the coagulation of manganese hydroxides — as sinters in cavities or massive.

In sedimentary deposits of manganese psilomelane occurs as compact intercalations or as oolites and globular concretions with a concentric-conchoidal structure. In the course of weathering psilomelane is oxidised and dehydrated, being converted into pyrolusite, mostly on the surface of cavities and pores, often as a black sooty mass. Among the biggest sedimentary manganese deposits containing psilomelane are those at *Chiatury* (Transcaucasia) and *Nikopol* (the Ukraine).

Uses: Together with pyrolusite and other manganese oxide minerals, it is the principal ore for ferromanganese. The low-manganese ores are added to blast-furnace charges.

2

Oxygen Salts

The complex compounds, from the chemical standpoint, are salts of various oxygen acids. This class includes nearly two-thirds of all the known minerals. Hence their importance in mineralogy.

Silicates are the most predominant among these components. Also very numerous are sulphates and phosphates. All of them occur in the earth's crust only in the solid state and are a product of chemical reactions proceeding under most diverse geological conditions. However, it should be remembered that salts are generally divided into *anhydrous* and *hydrous* (containing H_2O molecules), and further sub-divided as follows.

1. *Acid salts* (e.g., nahcolite or acid sodium carbonate, $NaHCO_3$) containing an H^{+1} proton instead of one metallic cation. These salts are comparatively rare in nature.
2. *Normal* or *medium* salts, such as calcite (calc-spar), $CaCO_3$, and gypsum, $CaSO_4.2H_2O$, which are the most widespread in nature.
3. *Basic* salts containing $[OH]^{1-}$ hydroxyl ions that neutralise the surplus positive charge of the cations.

They are very common in nature, e.g., malachite, $Cu_2[CO_3][OH]_2$, and aluminite, $Al_2[SO_4][OH]_4 \bullet 7H_2O$.

In the basic salts, the $[OH]^-$ anion, as has been confirmed in many instances, may be partly or completely replaced by other anions, mostly by the equidimensional F^- anion, sometimes by O^{2-} and Cl^- anions (the structure remaining unchanged). All such salts, including those containing OH, we shall term as *salts with additional anions.*

Binary and more complex salts, as compounds of definite composition, differ from simple salts in that either cations or anions, or both, are represented by different ions which do not or almost do not replace one another isomorphously. Examples: dolomite, $CaMg[CO_3]_2$, tychite, $Na_6Mg_2[CO_3]_4[SO_4]$.

Salts of *variable composition* reveal both isovalent and heterovalent isomorphism. In the latter case, positive and negative charges are always in equilibrium. Examples: ferrorhodochrosite, (Mn, Fe) CO_3; chillagite, Pb[MoO_4, WO]; and wilkeite, Ca_5[PO_4, SiO_4, SO_4] F, OH].

Crystallochemical features of compounds. The characteristic feature of the crystallochemistry of the oxysalts is the presence in their crystal structures of *anionic groups:* $[NOa]^-$, $[CO_3]^{2-}$, $[SO_4]^{2-}$, $[PO]^{3-}$, etc. The cations located in the centres of these groups have small ionic radii, high charges 206) and are linked with the oxygen ions by the covalent type of bonding.

It should be emphasised that the bonding of the oxygen ions with the central cation is very strong and depends on the so-called electrostatic valence, i.e., the share of the cationic charge taken by each surrounding oxygen ion. Thus, in the anionic group, which consists of one N^{5+} and three surrounding O^{2-} ions, each oxygen ion shares the charge in the ratio of 5 : 3; in the $[CO]^2$.

This ratio everywhere exceeds unity (i.e., more than half of oxygen valence is spent for the bonding with the central cation), and only for $[BO_3]^{3-}$ and $[SiO_4]^{4-}$ it is unity. This means that the oxygen ions are more strongly bonded with the smaller cations within the group than with the cations in the crystal structure outside the group. It is in this that the oxysalts differ materially from complex oxides for which the values of electrostatic valences of bonds between the cations and the surrounding oxygen ions are always less than unity.

Thus, anionic groups constitute very strongly bonded compact radicals participating in crystal structures as *independent* structural units. Their independent nature is also stressed by the fact that they do not disintegrate upon the dissolution of the salts. They differ materially from simple anions in shape and, naturally, in size. The JF_4-type anions have tetrahedral coordination and closely resemble isometric bodies in shape. The JF_3-type anions are flat triangles (CO_3 and BO_3), more seldom obtuse pyramids (AsO_3, etc.), in which the 0 ions are at the corners of the triangles, while the bonding cation is located over the centre of the triangle.

Apart from the anionic groups of simple configuration, there exist, as will be shown later, more complex radicals. It is noteworthy that they are found in borates and silicates, i.e., salts of the weakest acids.

The salts of stronger acids have crystal structures in which the type of bonding is typically ionic, as in halides, with the only difference that anionic groups take the place of simple anions. No wonder, therefore, that in a number of physical properties these salts closely resemble the minerals belonging to the class of chlorides and fluorides.

The principal anionic groups of simple structure, which occur in natural oxysalts. It should be noted that the concept of radii of anionic groups is a purely fictional one. The values cited were obtained by A. Y. Fersman by purely theoretical calculations and are all approximate. These tentative values are useful, however, for general consideration of the properties of minerals, especially when it is remembered that properties depending on the energy of the crystal lattice such as hardness, fusibility, volatility, relative solubility, etc., are estimated in practice by very rough scales.

Since the vast majority of oxysalts are typically ionic compounds, the stability of the crystal structures should largely depend on the ratio of sizes of the cations and anions as structural units. For the oxysalts, these ratios cannot be expressed in any numerical values

since the size of anionic groups cannot be expressed in the terms of radii, their configuration differing materially from the spherical shape which is conventionally attributed to simple ions.

However, there is no doubt that such isometric complex anions BiS $[SO_4]^{2-}$, $[PO_4]^{3-}$, and $[SiO_4]^{4-}$ should be far larger than simple anions (e.g., 0^{2-}). It is natural, therefore, that the anions in simple compounds of AX-type (where X is a complex anion) should give *the most stable* crystal structures only in combination with *large* cations. Such compounds indeed have low solubility, low fusibility, and low volatility. Here are some examples.

In compounds of the -AX-type (i.e., with anion-to-cation ratio of 1:1) the divalent $[SO_4]^{2-}$ anion gives the following most stable compounds (sulphates): barite, $BaSO_4$, and anglesite, $PbSO_4$, i.e., in combination with the *largest* divalent cations. This also explains the isomorphous admixture of Ra^{2+} in Ba^{2+} sometimes found in barites (the ionic radius of Ra^{2+} is somewhat larger than that of Ba^{2+}). No wonder, that in the oxidised zones of uranium deposits, which contain sulphides subjected to the action of sulphate waters, the concentration of radium in relation to uranium is higher than in the zones of primary ores (the compounds of hexavalent uranium are much more soluble under these circumstances than those of radium sulphate).

As to the divalent cations with *small* ionic radii in nature they are capable of forming only *hydrous sulphates* crystallising at low temperatures and at the last stages of crystallisation of solutions. In most instances, hydrous salts contain 2,4, 6 and 7 molecules of H_2O. X-ray studies of the $NiSO_46H_2O$ compound show that the Ni^{2+} cation in the crystal structure is surrounded by 6 electrically neutral (not compensating for the cationic charge) water molecules. Therefore, the volume of such hydrated cations is artificially increased, as it were, which makes possible a crystal structure containing such large anions as $[SO_4]^{2-}$. From the crystallochemical standpoint, the formula of this compound should be more correctly written as $[Ni(H_2O)_6]SO_4$. Such hydrous salts have definitely increased or high solubility and are easily dehydrated, their crystal structures undergoing rearrangement or complete destruction.

It should be noted that the medium-sized Ca^{2+} cation, which under certain conditions forms the $CaSO_4$ anhydrous sulphate (anhydrite), in the presence of water alters to the more stable hydrous $CaSO_4 2H_2O$ compound (gypsum). Gypsum, according to X-ray studies, has a layered structure in which the H_2O molecules are located between the sheets of Ca^{2+} and $[SO_4]^{2-}$ ions. Hence highly perfect cleavage of gypsum crystals in this direction.

The $[PO_4]^{3-}$ trivalent anion reveals a picture similar to that of sulphates. The largest trivalent cations, which are capable of combining with this anion into stable crystal structures, are the rare earths (Ce...) and La. Indeed, widespread in nature is a very stable rare-earth phosphate—(Ce, La...) PO_4 (monazite). In the process of weathering, monazite being a chemically inert mineral passes into placers. It is not accidental that the most important rare-earth concentrations are associated with this mineral. Another, less stable, phosphate, YPO_4 (xenotime) occurs less frequently. Phosphates with smaller trivalent cations (Al^{3+}, Fe^{3+}, Mn^{3+}) occur in nature as hydrous salts.

A significant fact is that the $[SiO_4]^{4-}$ anion forms the most stable compounds with large tetravalent cations (Zr^{4+}, arid Th^{4+}, often with an isomorphous admixture of U^{4+} uranium: zircon, $ZrSiO_4$, and thorite, $(Th,U)SiO_4$.

The crystal structures of thorite and its varieties are destroyed only through radioactive decay. Zircon, which does not contain radioactive elements, is a very stable mineral and therefore is widespread in the form of rolled grains in placers and sediments of most different ages. The bulk of zirconium in the earth's crust is concentrated in this mineral. Only in silica-poor and alkali-rich magmas is it capable of forming other minerals.

Classification. It is customary to classify salts according to their acid radicals or, which is the same thing, according to complex anions. The order in which they are placed may be the same as that of electrostatic valences or ionic potentials. It would be logical to begin the discussion of oxysalts, which immediately follow such oxides as the quartz group minerals, with silicates and borates crystallochemically closest to them and to finish it with the iodates, in the order of diminishing anionic potentials. However, for easier presentation, the reversed, simple-to-complex pattern seems more convenient, the carbonates coming under discussion before the sulphates.

The minerals corresponding to oxysalts fall into the following classes:

1. Nitrates.
2. Carbonates.
3. Sulphates.
4. Ghromates.
5. Molybdates and tungstates.
6. Phosphates, arsenates, and vanadates.
7. Borates.
8. Silicates.

Nitrates

In the earth's crust, nitrogen occurs exclusively in the form of complex anions such as $[NO_3]^{1-}$ and $[NH_4]^{1+}$, and in the form of gas in the atmosphere. The nitrates as salts of strong nitric acid, HNO_3, easily soluble in water are confined almost exclusively to recent formations in hot desert countries. They derive their nitrogen from the atmosphere. The oxidation reactions of the element are predominantly biogenic, i.e., associated with bacterial activity in soils or possibly with atmospheric electricity discharges (on high plateaus).

The most important are the nitrates of alkalis — K and Na. Less important are nitrates of alkali earths — Mg, Ca and Ba. In the oxidised zones of copper deposits, in the conditions of desert weathering, rare Cu-nitrates of complex composition sometimes occur. Compounds of nitrates with hydrates and salts of other acids are also known.

Soda Nitre, $NaNO_3$

Synonym: Chile saltpetre.

Chemical Composition: Na_2O 36.5 per cent, N_2O_5 63.5 per cent.

System, trigonal; *symmetry,* ditrigonal scalenohedral. Isostructural with calcite. Crystallises in rhombohedra which, in polar angle, are very close to the rhombohedra of calcite. Characteristically resembles calcite in a number of other properties as well. Twins are along (0112) and also along (0001), (0221) (1011). Generally as granular crusts and efflorescence.

Colour, white, grey, reddish brown, and lemon-yellow. *Lustre,* vitreous. $n_y = 1.585$.

Hardness, 1.5 to 2. Brittle. *Cleavage,* rhombohedral (1011), perfect. *Specific gravity,* 2.24 to 2.29.

Diagnostic Features: Gives a flash in the blowpipe flame on charcoal; fuses readily, colouring the flame yellow (Na). Readily soluble in water. Taste, slightly saline, cool.

Genesis: Formed in hot arid regions, devoid of vegetation, through biochemical decomposition (oxidation) of nitro gen-bearing organic matter (guano and other droppings of birds and animals) as well as of algae, nitrobacteria, etc. The rare atmospheric precipitations wash down newly formed nitre into hollows and ravines where soda nitre settles forming saltierras or even massive deposits.

Minerals occurring in paragenetic association with soda nitre are gypsum, mirabilite (sodium hydrosulphate), halite, occasionally iodates, etc.

The greatest deposits of soda nitre are located on a high plateau in *Chile*, at the foot of the Andes. For millions of years, beginning with the Upper Cretaceous period, the region was a very hot and arid desert (rain once in 4 or 5 years).

In the USSR, deposits of saltpetre, mainly potassium nitrate, are fairly widespread in the form of saltierra formations on hills and in ravines.

Uses: (1) saltpetre is the best of mineral fertilizers; (2) as a strong oxidiser, it is used in the treatment of nickel ores; (3) as decoloriser in the glass industry; (4) as a preserving agent for canned fish, meat, etc.; (5) as a component of gunpowder and various explosives.

With the advent of industrial catalytic synthesis of nitrogen compounds (ammonia and nitric acid) from atmospheric nitrogen, the world output of natural soda nitre declined sharply. However, the soda nitre deposits have retained their importance for the local industries.

Potassium Nitrate, KNO_3. Synonym: saltpetre.

System, orthorhombic; *symmetry*, orthorhombic dipyramidal. Space group *Pnma*. $a_0 = 5.42$; $b_Q = 9.17$; $c_0 = 6.45$. Occurs as friable white crusts and efflorescences similar to those of soda nitre. n_y = = 1.505. Hardness, 2. Cleavage, {011} perfect. Specific gravity, 1.99. Behaviour before the blowpipe is similar to that of soda nitre.

Potassium nitrate is more widespread than soda nitre. In the past India was an important producer (the deposits are now depleted). As crusts and efflorescences it is widespread mostly in the same localities as soda nitre.

Carbonates

This class includes numerous mineral species some of which are relatively widespread in nature. This is especially true of $CaCO_3$ which often forms enormous sedimentary series of marine origin. Carbonates are often accessories of metalliferous ore deposits and sometimes are a source of important metals, e.g., manganese (rhodo-chrosite), and iron (siderite).

The $[CO_3]^{2-}$ anion can form more or less stable compounds with cations of divalent metals with medium or large ionic radii. Such metals are few in number, the most important being Mg, Fe, Zn, Mn, Ca, Sr, Pb, Ba, and also Cu, Zn, Pb, etc., with additional $[OH]^{1-}$ and Cl^- anions.

Univalent cations (Na, K, and NH_4), which very closely approach these metal cations in size, may form anhydrous carbonates, provided, however, that the crystal structure contains an H^{1+} cation, i.e., in the form of acid salts. Hydrocarbonates are also characteristic of the Mg^{2+} cation, which is in general prone to hydration in aqueous environments.

Carbonates of trivalent metals are known for rare earths with an additional F^- anion. Of the smaller trivalent cations, the carbonates occasionally contain only Al^{3+}, and then always as binary and hydrous salts of divalent metals (Cu and Pb).

There are no carbonates of tetra- or pentavalent metals. Hexa-valent uranium forms a rare anhydrous carbonate with an additional O^{2-} anion, $UO_2(CO_3)$ (rutherfordite). All the other known carbonates of uranium are hydrous.

The carbonates possess certain specific physical properties. The hardness of anhydrous carbonates is never high, usually ranging from 3 to 5. They are well soluble in water. This is especially true of carbonates of alkalis and bicarbonates of those elements whose cations have relatively small ionic radii, e.g., Mg^{2+}, CO^{2+}, or on the contrary those, whose ionic radii are very large, e.g., Ba^{2+}. The Cu^{2+} carbonate occurs exclusively in the form of basic salts, which probably depends on the structure of the cation itself. This is also probably the cause of the intense green and blue colours of copper carbonates. All the other carbonates are either colourless or of pale colours. Optically the carbonates are characterised by comparatively high birefringence (n_x-n_z) due to the flat shape of the CO_3^{2-} anion.

Anhydrous and *hydrous* carbonates constitute the primary sub-divisions of the class, each being further divided into groups of isostructural chemical compounds. The remaining carbonates occur in nature as isolated mineral species and will be grouped according to their cations.

Anhydrous Carbonates

Calcite Group

This group comprises a great number of mineral species that are carbonates of the following divalent metals (in the order of increasing ionic radii): Mg, Zn, Fe, Mn, Ca, Sr, Pb, and Ba. It should be noted that ions with radii smaller than that of Ca yield extensive isomorphous series of minerals crystallising in the trigonal system, whereas ions with radii longer than that of Ca form carbonates in the orthorhombic system. The carbonate of calcium itself is dimorphous, i.e., it can crystallise in either system.

Carbonates........... Mg, Zn, Fe, Mu Ca

Trigonal system....... 0.74 0.83 0.80 0.91 1.04

Ca, Sr, Po, Ba

Orthorhombic system. 1.04, 1.20, 1.26, 1.38

If the cubic structure of NaCl is compressed along the threefold axis so that the interfacial angles become 101° 55', the result will be the rhombohedral face-centred structure of calcite 209), in which the Ca^{2+} ions will take the sites of Na, while the $[CO_3]$ groups— the sites of Cl. Thus the packing of ions in calcite corresponds to a somewhat distorted closest *cubical* packing of the structural units.

The crystal structure of aragonite, the orthorhombic $CaCO_3$ modification, differs from that of calcite in that the Ca^2+ and $[CO_3]^{2-}$ ions are arranged in the closest *hexagonal* packing 210). Hence the pseudohexagonal symmetry of crystal trillings (interfacial prism angles in some of the crystals are very close to 60 and 120°). Both in the calcite and aragonite structures each $[CO_3]^{2-}$ ion is surrounded by six calcium ions. Judging by the difference in specific gravity, the crystal structure of aragonite is more compact than that of calcite. Another important characteristic of the calcite minerals is the tendency to form isomorphous mixtures and binary salts. In the calcite series, Mg, Zn, and Fe extensively replace each other as it should have been expected in view of the relationship between the ionic radii of the cations.

The carbonates of $Fe^{2+}$and Mn^{2+} similarly form a continuous series of isomorphous mixtures. As for Ca, whose ionic radius differs materially from those of the above cations (except Mn^{2}+), it is capable of forming only binary salts with them. Calcium and magnesium ions (or other small cations) alternate in crystal structures parallel to the threefold axis. As a result, the symmetry of binary salts of the calcite series is of a somewhat lower order: instead of the ditrigonal scaleno-hedral class, we have a rhombohedral class, i.e., there are no twofold axes passing through the central carbon ion or the oxygen ions surrounding the latter.

Calcite: $CaCO_3$. Synonym: calcareous spar. It is a member of a very extensive isomorphous group. *Iceland spar,* a colourless and transparent variety has remarkable optical properties.

Chemical Composition: CaO—56 per cent, CO_2—44 per cent. Impurities: Mg, Fe, Mn (up to 8 per cent), much more rarely Zn (up to 2 per cent), Sr (strontian calcite), etc.

System, trigonal; *symmetry,* ditrigonal scalenohedral.

Crystal structure is typical of the group. Habit of crystals occurring exclusively in cavities is extremely diverse. The most common are scalenohedral crystals 211), more seldom tabular 212) or platy, prismatic or columnar; rhombohedrous, more frequently acute than obtuse. The known simple forms of crystals number several hundreds. The most common faces are prismatic {101.0}; rhombohedral {1011} (cleavage rhombohedron), {0112}, {0221}, {4041}; scalenohedral {2131}; pinacoidal {0001}, etc. The twinning plane is usually (0001) of a pinacoid 214) or the (0112) face of an obtuse rhombohedron, parallel to which polysynthetic twins are often formed in marbles and crumpled limestones (such twinning may be obtained artificially by exerting pressure with a knife edge on the cleavage edge of a calcite fragment). More rarely the twinning plane is the face of the cleavage rhombohedron (1011) 215), etc.

Aggregates: As stated earlier, druses of calcite crystals together with other minerals occur in cavities. Coarse-granular aggregates of transparent or translucent calcite, with individual granules possessing perfect cleavage occur fairly often. A rare occurrence is veined asbestos-like calcite with silky lustre (satin spar), the fibres of which run perpendicular to the walls of fissures in rocks. Well known are deposits of calcite in the form of stalactites and stalagmites in caverns in limestones. Large compact masses of calcite are known as *marbles*. Compact cryptocrystalline varieties of calcitic rocks often stratified and rich in fossil fauna, are called *limestones*. Friable limestones containing minute shells of foraminifera are known as chalk. There is also oolitic limestone sometimes also called roestone. The terms "calcareous tufa" and "travertine" are applied to the porous calcium-carbonate formations around cold and hot lime-saturated mineral springs (calcite in such cases forms through the recrystallisation of deposited $CaCO_3$ colloids, or aragonite). Sometimes, carbonate of lime depositing from hot springs produces thinly banded translucent compact varieties of a remarkable pattern, known as "onyx marble".

Colour, Predominantly colourless or milk-white, but when impure may have different (usually light) shades of grey, yellow, pink, red, brown, and black. Lustre, vitreous. $n_y = 1.658$ and $n_z = 1.486$. *Hardness*, 3. Brittle. Cleavage, {1011} perfect. Specific gravity, 2.6 to 2.8, and for chemically pure crystals 2.72 at 23°C. When compressed, yields twins and becomes electrically charged. Specimens from some deposits display luminescence, the cause of which is not clear yet.

Diagnostic Features: Coarse-crystalline varieties readily identifiable by rhombohedral cleavage, rather low hardness (can be scratched easily with a knife point or a needle), and by vigorous effervescence when the mineral or its powder is moistened with a drop of hydrochloric acid.

Decrepitates before the blowpipe flame with evolution of CO_2, i.e., dissociates into CaO and CO_2. The resultant CaO glows brightly and colours the flame orange. Dissolves well with effervescence in dilute hydrochloric acid (even cold) (evolution of CO_2).

Genesis: Calcite is one of the most widely distributed minerals in the earth's crust and may form whole massifs (limestone mountains). Its formation takes place in the course of most diverse geological processes.

1. Very widespread are crystalline formations of *hydrothermal* origin. It is formed in large masses in contact-metasomatic deposits through re-deposition or re-crystallisation of limestones. In pegmatites it is one of the last-formed minerals of the hydrothermal stage. Characteristically in numerous metalliferous, mostly sulphide vein deposits, calcite is, as a rule, among the last minerals to crystallise.

To the same class also belongs calcite amygdules and geodes in effusive igneous rocks, as well as aragonite and calcite deposited from certain mineral springs in the form of "calcareous tufa" (as a result of the vigorous evolution of CO_2 from the spring water due to abrupt decline of external pressure).

2. Although *weathering* processes do not give rise to large accumulations of calcite, some new formations are observed fairly frequently in fissures and cavities within oxidised zones of ore deposits and rocks.

In this case the source of calcite are atmospheric carbon dioxide and calcareous minerals decomposing in the weathering crust. It should be noted that in general, in the course of weathering of rocks large masses of lime pass into solution as calcium bicarbonate and may be transported by running waters over large distances, sometimes right to the sea, provided it does not meet on the way the conditions which cause it to precipitate as crystalline or colloidal $CaCO_3$ carbonate.

The stalactites in limestone caves result from the deposition of calcium carbonate from saturated solutions which seep slowly into the caves. Calcium carbonate is deposited when a hanging drop of the solution begins to evaporate, becomes strongly supersaturated, giving up a colloidal or finely-dispersed sinter-like precipitate which gradually solidifies and crystallises with further dehydration.

3. Enormous masses of $CaCO_3$ are formed by *sedimentation*, especially in marine basins, first as calcareous silts, dead marine plants and invertebrates with calcareous skeletons. With time all this material is converted to limestone. Oolitic limestone is most probably formed through coagulation of colloidal solutions of calcium carbonates around minute sand grains and gas bubbles suspended in turbulent water. Recent formation of oolites in marine basins is confined to shallow near-shore zones of tropical and subtropical seas. At first the oolites are composed of aragonite which later on is converted to calcite.

Irrespective of its mode of formation, calcite is a relatively unstable mineral in the weathering crust. In the eluvial zone, and especially in oxidised zones of sulphide-ore deposits, calcite is leached out because of its high solubility in acids. In the course of exchange decomposition reactions calcite is often replaced by other minerals (gypsum, dolomite, smithsonite, malachite, etc.) as shown below:

$$H_2SO_4 + CaCO_3 + H_2O \rightarrow CaSO_4 . 2H_2O + CO_2$$

$$ZnSO_4 + CaCO_3 + 2H_2O \rightarrow ZnCO_3 + CaSO_4 . 2H_2O, \text{ etc.}$$

Not only are coagulates of iron hydroxides, colloidal silica, etc., often observed on the surface of limestones in the zone where the ground water is saturated with calcium carbonate, but often the limestones themselves are completely replaced by the coagulates.

Large crystals of transparent calcite (Iceland spar) in the USSR have been recorded from deposits along the *Nizhnaya Tunguska River* through the recrystallisation of *de*ni effusive rocks—traps and mandelstones; here calcite occurs as veins and pockets containing gigantic monocrystal individuals. Numerous small deposits of Iceland spar are widespread in *Central Asia,* in the Zeravshan-Gissar, Pskem-Ugam, and other districts.

Notable foreign localities for Iceland spar are in *Iceland,* where it is found in effusive igneous rocks.

Chalk, as a rock, is widely distributed in thick series of Cretaceous carbonate rocks on the Russian platform and elsewhere in the Soviet Union. It is quarried on a large scale (to be used chiefly in chemical and cement industries) in the *Belgorod* district, in the vicinity of *Slavyansk, Kramatorsk* (Donets Coal Basin), and elsewhere.

Marbles of beautiful patterns are quarried in the Urals in the *Ufaley, Zlatoust* and other districts; in *Karelia,* the *Trans-Baikal area,* the *Crimea,* and elsewhere. Deposits of yellowish and greenish onyx marbles are known in Transcaucasia (Armenia and Georgia).

Famous foreign localities of excellent marbles are at *Carrara* on the eastern coast of the Gulf of Genoa (Italy) and in *Greece*-the wonderful antique statues were made of these marbles.

Uses: Calcite, especially calcite rocks, has many uses.

1. Iceland spar, owing to its high birefringence, is used in the manufacture of different polarising optical instruments, mainly for nicol prisms for microscopes, polarimeters, calorimeters, etc. For this purpose, colourless and perfectly transparent entwined crystals or fragments thereof are selected, which must have no cracks and be at least
1. 5 to 2 cm in size.
2. Asbestos-like satin spar and onyx marble are used as ornamental stones and gem materials.

3. Limestones, depending on their composition or physical and mechanical properties are used: (a) in chemical industry (pure limestones) for the manufacture of fertilizers, in the production of sugar, soda, caustic soda, chloride of lime, etc. (liquid or solid carbon dioxide is obtained simultaneously as a by-product of calcining); (b) in metallurgy (low-phosphorus and low-sulphur limestones) as flux in blastfurnace process; (c) in making slaked lime, Portland, Roman, and other cements; (d) in the printing industry as lithographic stones (compact cryptomerous limestone with a conchoidal fracture, readily splitting into thin slabs).
4. Polished marbles for facing the internal walls of buildings, for

statuary, and also in electrical engineering (switchboards, etc.).

5. Chalk is used for crayons and also in whiting, and polishing preparations as well as for the production of cement, paints, toothpowder, rubber (filler), etc.

Abagonite, CaCO3

Named after the place of discovery (Aragon, Spain).

Chemical Composition: The same as that of calcite: CaO- 56.0 per cent, CO_2- 44.0 per cent. Often contains impurities: Sr (up to 5.6 per cent), Mg, Fe, and Zn.

System, orthorhombic; *symmetry,* rhombic dipyramidal Crystal structure has been described earlier. Habit, prismatic, often pseudo-hexagonal, acicular. Principal forms: prism {110}, pinacoids {010} and {001}. The pinacoid may be striated parallel to the *a* axis. Other forms:

> Prism {011} giving the crystal a chisel-like termination, dipyramid {111}, sometimes very acute dipyramids combined with prisms, making the crystals spear-shaped). Twins on (110) are frequent as also are trillings of pseudohexagonal habit, fourlings and complex

polysynthetic twins. Therefore, groove-like reentrant angles between prism faces are quite common. Entwined crystals are extremely rare. Aggregates often occur as rod-like), radiated, and stellate intergrowths. Also observed as crystalline crusts, sinters, globular forms and oolitic masses (pisolitas or roestone). Very striking are occasional "iron flowers"—interlaced and branching "stems" of snow-white colour. And finally, the mother of pearl lining of most shells of molluscs is composed of very thin aragonite plates arranged parallel to the contours of the shell surface. As known, sand particles or any other foreign bodies penetrating inside the shell are coated with layers of calcium carbonate with some organic admixtures producing pearls.

Colour, of aragonite is white, yellowish white, sometimes lightgreen, violet, or grey. Some crystals are transparent and colourless.

Lustre, vitreous, greasy on fracture. In cathode rays shows weak pale-violet or orange luminescence n_x = 1.686, n_y = 1.681, and n_z = = 1.530.

Hardness, 3.5 to 4. Brittle. Cleavage, {010} distinct, {110} and {011} very indistinct. Fracture, often conchoidal. Specific gravity, 2.9 to 3.0 (higher than that of calcite, an evidence of closer packing of atoms). Other properties. Unstable at ordinary temperatures; in the presence of a solvent gradually changes to calcite, and therefore does not occur in older deposits. At 400°C, the change to calcite is very rapid. Curiously enough, those aragonite modifications that have not been polymorphously converted to calcite contain larger (up to several per cent) ions of strontium, which in the opinion of N. V. Belov, apparently has a stabilising effect on this modification. Aragonite is more soluble in water than calcite.

Diagnostic Features: Aragonite is very similar to calcite in colour and behaviour in hydrochloric acid but differs from it by the absence of rhombohedral cleavage, by the crystal habit with its characteristic, sometimes very fine, grooves on prism faces, and by higher hardness. Zeolites (hydrated silicates of Na, Ca, etc.) that appear to be similar to aragonite, do not evolve carbon dioxide in hydrochloric acid. Witherite and strontianite have a higher specific gravity and fuse before the blowpipe.

Before the blowpipe aragonite behaves like calcite. It is decomposed in acids with vigorous evolution of carbon dioxide. Powdered aragonite (as well as powdered strontianite and witherite) when boiled in a cobalt nitrate solution turns violet (Meigen's test), whereas powdered calcite suffers almost no change of colour, or turns bluish or greenish, and that only upon prolonged boiling.

Genesis and Occurrence: Aragonite is much less widely distributed in nature than calcite. Being one of the lowest-temperature minerals, it is fairly frequently formed towards the end of *hydrothermal* processes. Such is the origin, for instance, of aragonite found in fissures in serpentinised ultrabasic rocks where surface processes could have had no effect. To the same type of formation belong the small acicular aragonite crystals in the cavities of undecomposed basalts, occasionally in marbles, volcanic lavas, deposits from hot springs supersaturated with $CaCO_3$ in the form of calcareous tufa or oolites (pisolites of Karlovy Vary, Czechoslovakia), etc.

The bulk of aragonite is formed, however, by diverse *exogenous* processes, often with the participation of dissolved magnesial salts. As radial-fibrous formations and sinters which are quite often large, it is widely distributed in the weathering crust of ultrabasic magnesial igneous rocks in association with dolomite, gypsum, argillaceous matter, and other exogenous minerals. It also occurs in vugs in red iron ores as a crusty growths of small crystals and "iron flowers", as, *214*. Calcite twin on (0001). Figm 2*15*. Calcite twin on (1011).

Magnesite, $MgCO_3$: Magnesia is a region in Thessaly

(Greece). The mineral is known from ancient times. Synonym: magnesium carbonate.

Chemical Composition: MgO-47.6 per cent, CO_2-52.4 per cent. The most common isomorphous admixtures are Fe, sometimes Mn, Ca.

System, trigonal; *symmetry*, ditrigonal scalenohedral. Crystal structure, analogous to that of calcite. Habit, generally rhombohedral. Occurs mostly as coarse-granular aggregates. Porcelain-like metacolloidal masses, at times resembling cauliflower in shape, are very characteristic of deposits derived from weathering

Colour, white with a yellowish or greyish tinge, sometimes snow-white. Lustre, vitreous. $n_y = 1.700$ and $n_z = 1.509$. *Hardness*, 4 to 4.5. Brittle. *Cleavage*, rhombohedral on {1011} perfect. The fracture of metacolloidal porcelain-like formations is conchoidal. *Specific gravity*, 2.9 to 3.1.

Diagnostic Features: Crystalline varieties, like all carbonates of the calcite series, differs from other minerals by rhombohedral cleav-siderite occurs in typically *sedimentary* deposits formed in arine lagoons and bays. Their formation was evidently due to the producing conditions prevailing in the deeper parts of near-shore zones characterised by oxygen deficiency and by the decomposition of organ'; remains with the evolution of carbon dioxide and hydrogen sulphide from protein matter. Such sedimentary siderite ores sometimes have a typically oolitic texture.

Some of the ores of the great *Kerch* sedimentary deposit are composed of siderite. It is possible that brown iron ore is partly the product of oxidation of these ores. Outside the Soviet Union, notable are the rather large deposits of compact argillaceous spherosiderite, coloured dark due to admixed carbonaceous matter found in extensive coal beds in *Scotland* and *South Wales* (Britain).

In the oxidised zone siderite is unstable readily forming iron hats Composed of limonite, goethite, sometimes hydrohaematite and occurring as loose, generally earthy masses and as geodes, which are often yellow inside.

Uses: When siderite occurs in large masses with little undesirable impurities (phosphorus, sulphur, etc.), it is an ore of iron. Prior to Belting, siderite ores are calcined.

Rhodochrosite, $MnCO_3$; "Rhodon" is the Greek for rose; "chros", colour. The name is descriptive of its colour. Synonym: dialogite.

Chemical Composition: MnO 61.7 per cent (Mn 47.8 per cent), CO_2 38.3 per cent. The most frequent isomorphous admixtures are Fe, Mg, Ga, occasionally Zn and Co.

System, trigonal; *symmetry,* di-trigonal scalenohedral.

Crystal structure analogous to that of calcite. Well-formed Crystals are rare and occur only in cavities. Common forms: {1011} and {0112}, sometimes {0001} and {1120}. Faces are often saddle-shaped or lenticularly curved. Aggregates are commonly crystalline-granular, reniform and globular with a radiated or spherulitic texture. Occurs also in rod-like aggregates and earthy masses, colour of crystals pink or rasp. With increasing calcium content colouring pales. On exposure to air gradually turns brown (oxidisation). Fine-granular and earthy masses have a white colour with a very slight pink tinge. Streak, white. Lustre, vitreous. n_y = 1.817 and n_z = 1.597. *Hardness,* 3.5 to 4.5. Brittle. *Cleavage,* rhombohedral parallel {1011}. *Specific gravity,* 3.6 to 3.7.

Diagnostic Features: In crystalline masses rhodochrosite is readily identifiable by rhombohedral cleavage, pink colour, and low hardness (is scratched with a knife blade). White cryptocrystalline and earthy aggregates may be identified positively only by chemical and spectral analysis.

Infusible in the blowpipe flame but decrepitates, and successively turns greenish grey and black (owing to oxidation). With borax and salt of phosphorus gives the manganese reaction (in the oxidising flame gives a violet bead, a colourless one in the reducing flame. Dissolves slowly in cold hydrochloric acid, but when heated the dissolution proceeds rapidly with effervescent evolution of carbon dioxide.

Genesis and Occurrence: Judging by the composition of the iron-manganese carbonates, $MnCO_3$ and $FeCO_3$ give a continuous series of isomorphous mixtures. The dark of manganese in the earth's crust is about 50 times lower than that of iron. Nevertheless, in nature rhodochrosite forms deposits of its own.

In rare *hydrothermal* vein or metasomatic deposits of manganese, rhodochrosite is formed in association with sulphides and silicates of manganous oxide, crystallising after braunite, quartz, barite, etc. Rhodochrosite of hydrothermal origin in association with pyrite, chlorites, and other minerals is found in the USSR, in the *Sapalsk* deposit in marbled limestones near Nizhny Tagil. In certain tungsten deposits, rhodochrosite may occur as veins genetically associated with quartz veins containing hiibnerite ($MnWO_4$) and other minerals.

In much greater masses rhodochrosite occurs in *sedimentary* deposits of manganese ores. According to the data of geological research, opal-rhodochrosite sediments have formed at some distance from the coastline in the deeper parts of the basins where the decomposition of organic remains in the conditions of oxygen deficiency evidently produced reducing conditions. In such deposits, rhodochrosite commonly contains isomorphous admixtures of Ca, Fe, and Mg and is associated with iron sulphides, manganocalcite, opal, etc. As a rule, sedimentary carbonate manganese ores are also enriched in phosphorus. Considerable masses of rhodochrosite ores are known in the sedimentary deposits of *Chiatury* (Transcaucasia), *Polunochny* (the Northern Urals), and elsewhere.

Uses: Low-phosphorus hydrothermal rhodochrosite ores are an important material for ferromanganese. Sedimentary carbonate ores may be used as an addition to blast-furnace charges and also by the chemical industry.

Smithsonite, $ZnCO_3$: Synonym: zinc spar.

Chemical Composition: ZnO 64.8 per cent (Zn 52 per cent), CO_2-35.2 per cent. Commonly contains isomorphous admixture of Fe, Mn, Mg, sometimes Co, rarely Cd, In, etc.

System, trigonal; *symmetry*, ditrigonal scalenohedral. Crystal structure, the same as of calcite. Rare crystals have both rhombohedral and scalenohedral habit. The most frequent forms: rhombohedral {1011} and {4011}; scalenohedral {2131}; more seldom pinacoidal {0001}; rhombohedral {0112}, {0221}; and prismatic {1120}. Faces are often curved and rough. Aggregates. Usually in earthy or compact crypto-crystalline aggregates, often as sinters and crusts as well as conchoidal 223), cellular, and porous masses.

Colour, white with greenish, brownish, and greyish tinges. Intensely green varieties contain malachite admixtures. Brown varieties owe their colour to iron hydroxides. Lustre, vitreous. n_y = 1.849 and n_z = 1.621. In cathode rays shows a faint pink luminescence. *Hardness*, 5 (the highest for the calcite group minerals). Cleavage, rhombohedral, noticeable only in crystalline aggregates. Specific gravity, 4.1 to 4.5.

Diagnostic Features: When in crusts, sinters and cryptocrystalline masses, is distinguished with difficulty from its paragenetic associates such as opal, zinc silicates (calamine, willemite), etc. Being rather light in colour, it is inconspicuous among limestones. Identification always requires blowpipe and chemical analysis. Other less important characteristics of smithsonite are its high specific gravity and hardness.

Infusible in the blowpipe flame and, like all the carbonates, decrepitates, and on charcoal gives a white coating of ZnO. Ferruginous varieties become brown. Rather readily soluble in acids, sometimes with effervescence (especially the earthy varieties). Upon ignition moistened in a $Co[NO_3]_2$ solution and ignited once more in the oxidising flame, becomes green.

Genesis and Occurrence: Smithsonite is a typical mineral of the lower horizons of the oxidised zones of lead-zinc sulphide *deposits in limestones.* Sometimes it forms large masses, mostly in the foot walls of ore bodies. When the primary ores contain much calcite, smithsonite is also found in the upper parts of oxidised zones, associated with zinc silicates, galena, and sometimes with remains of sphalerite. It is formed metasomatically through the exchange-decomposition reaction of well-soluble zinc sulphate with calcite:

This reaction probably occurs after the neutralisation of the excess of sulphuric acid in solution through reaction with the same calcite. If *ferrous* oxide sulphate is present in the solution, which may happen only in the deeper horizons of the oxidised zone, smithsonite contains isomorphous admixture of $FeCO_3$ sometimes in large amounts (mon-heimite). Ferruginous smithsonite decomposes under oxidising conditions with separation of iron hydroxides. In this case the zinc in the form of smithsonite which has already been stripped of iron, may be re-deposited elsewhere. Smithsonite masses intensely pigmented with iron hydroxides differ little from porous or cavernous masses of limonite, and are therefore frequently overlooked. The smithsonite may be recognised by blowpipe analysis or in polished sections under the microscope.

In the USSR, smithsonite in large masses occurred at *Turlan* (Achisai); in the Kara-Tau Ridge (Southern Kazakhstan), which was for a long time thought to be a purely lead deposit, until rich oxidised zinc ores were discovered beyond the principal ore body); in numerous lead-zinc deposits of the eastern Trans-Baikal area (Ner chinsk district), arid elsewhere.

Outside the USSR, there are numerous deposits, among which the most important is at *Leadville,* Colorado (U.S.A.). At first, it was worked in the oxidation zone only for its lead; only 30 years later rich smithsonite ores were found here (so inconspicuous they were in limestones and other rocks).

Uses: Smithsonite ores, if found in quantity, may be an important source of zinc. Their zinc content is often 2 to 3 times higher than in-primary sulphide ores. Thus, in the lower parts of oxidised zones of lead-zinc sulphide ores in limestones or dolomites, not only zinc is separated from lead but the metasomatically deposited exogenous products are also enriched in Zinc.

Cerussite, $PbCO_3$: "Cerussa" is the Latin for white paint. Synonym: white lead ore.

Chemical Composition: PbO 83.5 per cent (Pb 77.5 per cent), CO_2 16.5 per cent. May contain as mechanical impurities dispersed residual PbS and Ag_3S, colouring the mineral black, occasionally $ZnCO_2$.

System, orthorhombic; *symmetry,* rhombic dipyramidal. Crystal structure, analogous to that of aragonite. Habit may be diverse, viz., pseudohexagonal dipyramidal with dipyramid faces {111} and prism faces {021}; tabular or platy with predominant pinacoidal development {001} or {010}, sometimes {100}; rod-like, etc.

Twins and trillings are frequent; the most common twinning plane {110}. Aggregates usually granular massive. Less often sinter-like, cryptocrystalline and earthy masses. Also occurs in snow-white fibrous varieties.

Colour, commonly white with greyish, yellowish, or brownish tinges. Ferruginised masses are brown and therefore often overlooked. Occasionally grains are black due to microscopic inclusions of residual sulphides. Individual crystals are often colourless or transparent. Lustre, adamantine, sometimes vitreous (depending on orientation).

Fracture, of ten uneven, conchoidal. In cathode rays shows bright greenish-blue fluorescence. $n_x = 2.078$, $n_y = 2.076$, and $n_z = 1.804$.

Hardness, 3 to 3.5. Very brittle. Cleavage, sometimes observed on {110} and {021} distinct. Specific gravity, 6.4 to 6.6.

Diagnostic Features: Unlike the other carbonates, cerussite has high specific gravity and adamantine lustre. Often observed in association with anglesite and galena.

Decrepitates strongly before the blowpipe, turns yellow (PbO), on charcoal is reduced readily to metallic lead. Dissolves in dilute nitric acid with brisk evolution of carbon dioxide. Also soluble in KOH.

It is interesting to note that cerussite masses coloured with iron hydroxides are externally indistinguishable from other limonitised materials containing no cerussite. However, when broken, a piece of such cerussite emits a characteristic crackle or squeak. Miners working the cerussite portions of the oxidised zones of ore deposits rely on this property to distinguish cerussite.

Genesis and Occurrence: Cerussite is confined almost exclusively to oxidised zones of lead-zinc sulphide deposits. Usually occurs as an alteration product of anglesite which, in turn, is formed through the oxidation of galena. Being almost insoluble in water and stable in air, cerussite protects galena from further decomposition.

Well-formed, and often large, crystals of cerussite occur on the walls of leaching cavities in the oxidised zones. The formation of these crystal druses implies that there has been some transportation of soluble lead compounds. However a possibility of formation of cerussite at the low-temperature stage of hydrothermal processes is not excluded.

Pseudomorphs of cerussite after other minerals (galena, anglesite, calcite, fluorite, etc., are comparatively rare. Large masses of cerussite were mined in the USSR in the *Turlan* deposit, Kara-Tau Ridge (Southern Kazakhstan), and elsewhere. Extremely well-formed cerussite crystals occurred at the Tayninskoye and Kadainskoye deposits in the *Nerchinsk* district, Trans-Baikal area, and in the *Altai mountains* (Ridder, Zyryanovskoye and Nikolayevskoye).

Uses: An important lead ore, especially in extensive oxidised zones of lead-zinc deposits.

Strontianite, $SrCO_3$

First discovered at Strontian (Western Scotland).

Chemical Composition: SrO-70.2 per cent, CO_2 -29.8 per cent. Practically always contains CaO. The variety known as *calciostrontianite* contains up to 13 per cent of CaO. Less often contains BaO, PbO, etc.

System, orthorhombic; *symmetry,* rhombic dipyramidal. At a temperature of 700°C (according to other data, at 929°C), changes to a hexagonal modification which never occurs in nature. Crystal structure, analogous to that of aragonite. Crystals are rare, mostly as thin needles or barrel-shaped prismatic forms. Twins are common along (110). Aggregates usually massive granular, often thin rod-like or fibrous.

Colour, Colourless or greenish, yellowish, and greyish. *Lustre,* vitreous; on fracture, greasy. In cathode rays gives very faint bluish fluorescence. $n_x = 1.668$, $n_y = 1.667$, and $n_z = 1.520$.

Hardness, 3.5 to 4. Brittle. *Cleavage,* {110} distinct and {021} indistinct. Specific gravity, 3.6 to 3.8.

Diagnostic Features: Externally, hard to distinguish from aragonite. Behaviour before the blowpipe and reactions with acids are highly characteristic.

Before the blowpipe, when intensely heated, swells into cauliflower shapes. Incandesces strongly and colours the flame intense carmine-red (Sr), readily soluble in acids with effervescence. When its solution in hydrochloric acid is evaporated and some alcohol poured on the residue a bright-red flame (Sr) flares up.

Genesis and Occurrence: Occurs mostly in *hydrothermal* formations in association with celestite, barite, calcite, sulphides, and other minerals, also in sedimentary rocks (limestones and marls), although frequently in the form of more recent veinlets in cavities and cracks.

In the USSR, strontianite deposits occurring the Crimea. Outside the Soviet Union, large deposits of strontianite were worked at *Hamm,* Westphalia (Germany) where they occur as veinlets and veins in Cretaceous marls. The cavities there contained very well developed crystals. As an accessory mineral, strontianite occurs in many hydrothermal deposits, mostly in barite veins accompanied by sulphides.

Uses: A secondary source of strontium.

Witherite, $BaCO_3$: Chemical composition. BaO 77.7 per cent, CO_2 22.3 per cent. Sometimes contains strontium.

System, orthorhombic; *symmetry,* rhombic dipyramidal. Crystal structure, analogous to that of aragonite. Crystals occur as pseudo-hexagonal dipyramids. Trillings are common on (110). Aggregates. Also occurs in globular and reniform shapes, sometimes in veined, fibrous, and foliated aggregates.

Colour, Colourless or white, but usually coloured greyish or yellowish. Lustre, vitreous; Fracture, greasy. In cathode rays, shows yellow fluorescence. n_x = 1.677; n_y = 1.676, and n_z = 1.529.

Hardness, 3 to 3.5. Brittle. *Cleavage,* {010} distinct and {110} indistinct. *Specific gravity,*

4.2 to 4.3. Other properties. Witherite dust is toxic to man. Therefore, the wet method is *228.* Witherite. Penetration trilling sed in drilling blastholes in witherite-bearing ores and rocks.

Diagnostic Features: High specific gravity distinguishes witherite from otherwise similar aragonite and strontianite. Also characteristic are its blowpipe behaviour and reactions with acids.

Before the blowpipe fuses readily into transparent bead which looks like enamel on cooling. This feature places it apart from all other carbonates. The flame is coloured characteristically yellow greenish. Soluble with effervescence in dilute nitric and hydrochloric acids. On addition of a few drops of sulphuric acid, an abundant precipitate of $BaSO_4$ falls out.

Genesis and Occurrence: Usually occurs in *hydrothermal* deposits in paragenesis with calcite, dolomite, sulphides of Pb, Zn, Fe, and often with barite which is sometimes deposited after witherite. May occur as an exogenous mineral. Occurs in pseudomorphs after barite, formed apparently, through the action of carbonate solutions. Replacement of witherite by barite has also been observed.

Large accumulations of witherite are extremely rare. The one at *Settlingstone* in Northumberland (Britain) is actually the only specific witherite deposit. Witherite there is accompanied by calcite, sulphides, and later-formed barite (in thin cracks and in the spaces between witherite crystals). In the Soviet Union, barite-witherite deposits occur as a series of veins in the Karakal district (Turkmenia) where they are confined to the fissures of a large tectonic fractured zone in Cretaceous sedimentary rocks.

Uses: Witherite is a secondary source of barium compounds for the reason that it occurs in nature much more rarely than barite. The uses of barium-bearing minerals are described under "Barite" (p. 372).

Parisite, Ca (Ce, La.. .)2[CO_3]$_3$F

Contains up to 50 per cent of $(Ce, La)_2 O_3$. A very rare mineral.

System, hexagonal. Usually occurs in compact granular masses.

Colour, reddish brown, brownish yellow. *Lustre,* vitreous. n_x = 1.757 and n_y = 1.676. *Hardness,* 4 to 4.5. Cleavage or parting, {0001} distinct. *Specific gravity,* 4.35. Infusible in the blowpipe flame. Decomposes slowly in hydrochloric acid. When heated with sulphuric acid, HF is evolved.

Occurs in pegmatites. Finds have been recorded from emerald mines in Muso, Columbia (South America), and elsewhere.

Summary: Thus, the calcite-group minerals have many physical properties in common. It will be seen from the general table of main properties of the principal calcite minerals, that only the carbonates of heavy metals such as Mn, Fe, and Zn are somewhat different from the rest, especially to the carbonate of lead cerussite. The greatest differences are in the solubility of the metals in water saturated with carbon dioxide. The carbonate of Mg is highly soluble; carbonates of Ba and Ca come next, and the carbonate of Pb is the least soluble of all.

Basic Properties of Calcite Minerals

Mineral	System	Specific gravity	Hardness	Refractive index	(at 180° and CO_2 pressure of 1 atm), mg	Solubility		Dissociation
				n_g	n_t	per litre		n_t
Calcite, $CaCO_3$	Trigonal	2.7	3	1.686	1.66	1.486	1,100	900
Dolomite, $CaMg[CO_3]_2$	Ditto	2.9	3.5-4		1.68	1.5	25,800	600
Magnesite, $MgCO_3$	Ditto	3	4 -4.5	1.677	1.7	1.509	400	500
Rhodochrosite $MnCO_3$	Ditto	3.7	3.5-4.5	1.667	1.82	1.597	720	500
Siderite $FeCO_3$	Ditto	3.9	3.5-4.5	2.078	1.88	1.633	700	400
Smithsonite $ZnCO_3$	Ditto	4.4	5		1.85	1.621	2,700	900
Aragonite $CaCO_3$	Orthorhombic	3	3.5-4		1.68	1.53	1,250	800
Witherite $BaCO_3$	Ditto	4.3	3 -3.5		1.68	1.529	140	800
Strontianite, $SrCO_3$	Ditto	3.7	3.5-4		1.67	1.52		
Cerussite $PbCO_3$	Ditto	6.5	3 -3.5		2.08	1.804		

Which make up isomorphous series, their properties vary depending on the ratio between their constituents. The same applies to binary salts. The thermal dissociation of the latter proceeds in two stages.

Malachite Group

Included in the group are basic anhydrous copper carbonates—malachite and azurite.

Malachite, $Cu_2[CO]_3[OH]_2$ or $CuC93.Cu[OH]_2$

"Malache" is the Greek for hollyhock. Apparently derives its name from the likeness of its colour to that of the plant.

Chemical Composition: CuO 71.9 per cent (Cu 57.4 per cent), CO_2 19.9 per cent, H_2O 8.2 per cent. May contain very small amounts of CaO, Fe_2O_3, SiO_2, etc., apparently as adsorbed or mechanical foreign admixtures.

System, monoclinic; *symmetry,* prismatic. Space group a_0 = 9.49; b_0 = 12.00; c_0= 3.24; *0* = 98°42'. Extremely rare crystals are of prismatic habit.

Aggregates usually in rounded masses with a radial-fibrous structure. Large reniform masses have a characteristic concentric banded structure, most striking in polished specimens. Also occurs in earthy varieties (copper green).

Colour, green. *Streak,* pale-green. *Lustre,* vitreous to adamantine, and for fibrous varieties, silky. n_x = 1.909,/$_{ny}$ = 1.875, and n_z = 1.655.

Hardness, 3.5 to 4. Brittle. *Cleavage,* {201} perfect and {010} distinct. *Specific gravity,* 3.9 to 4.0.

Diagnostic Features: Readily identifiable by green colour, frequent sinter forms, and radially fibrous structure. Differs from similar chrysocolla (copper hydrosilicate), phosphorohalcite (copper phosphate), and other green copper minerals by its behaviour in hydrochloric acid (evolves carbon dioxide). Fuses in the reducing blowpipe flame and gives a copper globule. When moistened with hydrochloric acid, colours the flame light-blue. Gives water in the closed tube and turns black. Soluble with effervescence in hydrochloric acid. The solution turns blue when ammonia is added.

Genesis and Occurrence: Malachite forms exclusively in the oxidised zones of copper sulphide deposits, especially in limestones or when the primary ores contain much carbonates. The most common copper mineral in oxidised copper ores, malachite develops either through the replacement of carbonates or by filling cavities as typical colloform (sinter) forms. Since at the surface of limestones or Ca and Mg carbonates solutions become definitely alkaline, the copper sulphate solutions on reaching this environment are hydrolised by reacting with bicarbonate solutions.

Moreover, copper carbonates may probably derive from the slow reaction of a copper sulphate or hydrate with a solution saturated with atmospheric carbon dioxide.

Malachite occurs frequently as pseudomorphs after azurite, cuprite, native copper and also after such minerals as atacamite, calcite, chalcopyrite, etc. As inclusions and salvages of copper green frequently occurs in the oxidised zones of copper deposits. However, large accumulations, especially those of ornamental grade, have now become a rarity.

In the USSR *Mednorudyanskoye* and *Gumeshevskoye* deposits were of world renown.

The Mednorudyanskoye was the only deposit in the world that supplied malachite of rich and beautiful patterns. It is from there that came the facings of the columns of St. Isaac's Cathedral (Leningrad), of the Malachite Hall of the Winter Palace, the malachite table tops of the Hermitage, etc. The first large block of malachite weighing about 50 tons was found at a depth of 70 m together with other oxygen compounds of copper, in strongly ferruginised and decomposed rocks at the contact between skarns and limestone.

The Gumeshevskoye deposit was discovered in the late

18th century. It had supplied an enormous quantity of excellent ornamental-grade malachite, specimens of which were displayed in many museums. The large sinter masses at that deposit occurred in red clays. It is from here that came the enormous block of malachite, of most beautiful pattern, weighing almost 1.5 tons, displayed at the museum of the Leningrad Mining Institute.

Uses: Compact sinter varieties of malachite, sometimes occurring in large masses, are used for facings and mosaics, vases, trinket-boxes and other ornamental objects. Malachite fines are used in the manufacture of pigments. Disseminated earthy malachite, along with other oxidised ores of copper, is a source of metallic copper.

Azurite, $Cu_3[CO_3]_2[OH]_2$ or $2CuCO_3$= $Cu[OH]_2$

The name comes from "azure "meaning sky-blue. Synonyms: azure copper ore, chessylite. Chemical composition. CuO 69.2 per cent (Cu 55.3 per cent), CO_2 25.6 per cent, H_2O 5.2 per cent. Crystals are chemically rather pure. A rare case of mechanical impurity was an admixture of specks of secondary native gold in some specimens from the Berezovskoye deposit in the Urals.

System, monoclinic; *symmetry,* prismatic. Crystals occur as short columns or prisms and also as thick tablets or plates. Often in druses of fine crystals, continuous granular masses, sometimes in radial and earthy aggregates ("blue malachite").

Colour, dark blue, in earthy masses, azure. *Streak,* light blue. *Lustre,* vitreous. n_x = 1.838, n_y = 1.758, and n_z = 1.730. *Hardness,* 3.5 to 4. Brittle. *Cleavage,* {001} perfect and {100} indistinct. *Specific gravity,* 3.7 to 3.9.

Diagnostic Features: Readily identifiable by °*a* {100}' characteristic blue colour and association with mala*m* {Yio}, *x* {112}', chite and other oxygen copper compounds and *d* {123}. Readily fusible in the blowpipe flame; in the reducing flame gives a copper globule. Soluble in acids with effervescence. Also dissolves in ammonia, colouring the solution light blue.

Genesis: In small amounts, occurs practically always in paragenesis with malachite, frequently being deposited after it. Conditions often arise under which azurite is less stable than malachite and hence is replaced by the latter. Malachite pseudomorphs after fine azurite crystals are known. Characteristically, such malachite occurs as continuous cryptocrystalline masses instead of having a radial-fibrous structure.

Uses: With other oxygen copper compounds goes into the smelts. Pure azurite, when found in quantity, is used for the manufacture of blue pigment.

HYDROUS CARBONATES

The most common of the carbonates containing H_2O molecules, the hydrous carbonates of Na, Mg, U, etc. Only dehydrocarbonate of sodium will be discussed here.

SODA, $Na_2CO_3.10H_2O$

The name "soda" was used already in the 17th century, though its origin is unknown. Synonyms: natrite, natron.

Chemical Composition: Na_2O 21.6 per cent, CO_2 15.4 per cent, H_2O 63.0 per cent. Mechanical impurities may consist of other well-soluble sodium salts.

System, monoclinic; *symmetry,* prismatic. Crystals are rhombohedral tabular. Usually occurs in granular aggregates.

Colour, white, grey or colourless. *Lustre,* vitreous. n_x = 1.440, n_y = = 1. 425, and n_z = 1.405.

Hardness, 1 to 1.5. *Cleavage,* {100} perfect. *Specific gravity,* 1.42 to 1.47. Precipitates from pure saturated solutions of Na_2CO_3 under atmospheric pressure, in the range of temperatures between —2 and + 32°C.

Diagnostic Features: Readily soluble in water. When treated with hydrochloric acid vigorously evolves carbon dioxide. On exposure to air, rapidly loses water and turns white. When slightly heated, melts or rather dissolves in its own crystallisation water, like many other water-abundant crystallohydrates, with separation of thermonatrite ($Na_2CO_3.H_2O$). When the precipitate is fused in a platinum loop, the flame becomes intense yellow (sodium).

Genesis: Is formed in quantity in certain saline sodium-rich lakes, with excess CO_2. Presumably the soda derives from the exchange decomposition reaction between Na_2SO_4 and inflowing calcium bicarbonate, $Ca[HCO_3]_2$. It may also be formed biochemically through the reduction of sulphates by microorganisms and subsequent action of carbon dioxide with evolution of H_2S, according to the reactions:

$$Na_2SO_4 \rightarrow Na_2S + 2O_2$$

$$Na_2S + H_2O + CO_2 \rightarrow Na_2CO_3 + H_2S$$

As snow-white efflorescences and coatings, soda appears on the surface of loose rocks and soils in countries with an arid and hot climate.

In the Soviet Union, all the large soda lakes are situated in the Asiatic part of the country: the Petukhovskoye and Mikhailovskoye lakes in the *Kulunda Steppe,* south of the Kulunda railway station (Northern Kazakhstan). The deposition of soda begins in autumn with decreasing temperature and it is recovered from the bottom in winter.

Thus, in the *Doroninskoye* soda lake (Eastern Siberia) the brine is pumped onto the ice in a layer up to 6 cm deep. The soda from the freezing brine precipitates on the ice of the lake. Upon the evaporation of the water of the frozen brine by strong winds, the soda is raked into heaps and delivered to the factory for treatment.

Numerous large soda lakes are known in Tripoli, Egypt, Iran, Tibet, southeastern California, and elsewhere.

Uses: Soda is used in diverse industries such as soap-making, glass-manufacture, dyeing, and also in metallurgy and chemistry. In countries where there are few soda lakes, soda is made artificially, chiefly from mirabilite and halite.

SULPHATES

The geochemistry of sulphur, it will be recalled, has a number of specific features. This element capable of forming riot only neutral 8-atom S_8 molecules, but also differently charged positive and negative ions. It will also be remembered that there exist negatively charged S^{2-} ions (analogous to O^{2-}) and $[S_2]^{2-}$ ions—as the products of electrolytic dissociation of H_2S. The formation of sulphides is associated with these anions. In an oxidising environment, sulphur may form an $[SO_3]^{2-}$ molecular compound (sulphurous gas) and, in solutions — $[SO_3]^{2-}$ complex anions and in a still more oxidising environment — $[SO_4]^{2-}$ groups containing S^{4+} and S^{6+} cations. Crystalline formations constituting the compounds of metals with these anions are known as sulphites (never occur in nature) and sulphates (widely occur in nature).

Thus, the formation of metal sulphates occurs only under conditions of increased oxygen concentration, i.e., at increased partial oxygen pressure in the environment, and at relatively low temperatures. In the earth's crust such conditions occur near the surface where the vast majority of these compounds, both endogenous and exogenous, are found.

Although the minerals of this class include a wide variety of compounds, the number of stable and widely occurring minerals is rather limited.

So large a complex anion as the $[SO_4]^{2-}$ may form stable crystalline structures only in combination with large divalent cations. Indeed, as will be shown later, the most stable are the sulphates of Ba, Sr, and Pb. Cations with smaller ionic radii enter into the composition of sulphates only in a hydrated state, i.e., when clad in an H_2O molecular "jacket". The smaller the cation the greater is the number of H_2O molecules that can be attached to it. In the case of water-rich sulphates, the amount of H_2O in the same type of compound may vary depending on the H_2O vapour pressure in the environment. Thus, for the sulphate of ferrous oxide there exist salts with seven, six, five, and one H_2O molecules. All these salts differ from one another in crystal lattice.

Univalent cations of alkali metals naturally enter into the composition of simple sulphates either in double numbers or in association with H^{1+}. They form loose crystal structures and are easily soluble in water (as are sulphates with smaller divalent cations).

Sulphates of trivalent metals, mainly of Al^{3+}, and Fe^{3+} occur exclusively as hydrous compounds.

Very widespread are binary and more complex salts of uni-, di-, and trivalent metals. Very frequent are sulphates with additional an-ions such as $[OH]^{1-}$, occasionally with Cl^{1-}, $[CO_3]^2$, $[PO_4]^{3-}$, etc., especially when they contain trivalent metals or strongly polarising Cu^{2+} cations.

In conclusion we shall note certain physical properties common to all the minerals of this class, the most striking of which is their low hardness. There are no sulphates with a hardness over 3.5 and for water-abundant varieties it is as low as 2. Comparing the optical properties of the sulphates with those of the earlier reviewed salts, we find that their birefringence *(n_x-riy)* is much lower.

Some sulphates are even optically isotropic, which is explained by the fact that as different from the flat GO_3 and NO_3 groups, the tetrahedral SO_4 groups are isometric structural units. Most characteristically sulphates containing additional $[GO_3]^{2-}$ anions have considerably higher birefringence (e.g., cale-donite, Pb_5Gu_2 $[SO_4]_3$ $[GO_3][OH]_6$.

In this description the minerals will be divided into groups, as usual, according to chemical composition and crystal structures. However, few such groups are known for anhydrous minerals.

Other, rather numerous minerals of diverse composition, are best grouped according to cationic valency, and therefore hydrous and anhydrous sulphates will be discussed together.

BABITE GROUP

The group includes those sulphurous compounds of Sr, Ba and Pb which do not occur in nature as hydrosulphates. Anhydrous sulphate of calcium (anhydrite) although crystallising in the orthorhombic system, as the above sulphates, differs from the both in crystal structure and crystal form, since the Ca^{2+} ion has a smaller radius. Since anhydrite is genetically very closely associated with hydrous calcium sulphate (gypsum), it will be discussed in detail together with gypsum.

Babite

$BaSO_4$ "Baros" is the Greek for heavy. The high specific gravity of the mineral is felt even by the hand. Barite is the principal of the very few barium minerals, and the most widespread anhydrous sulphate after anhydrite.

Chemical Composition: BaO 65.7 per cent SO_3 34.3 per cent-Sr and Ca are found sometimes as isomorphous admixtures. The variety with a high strontium content is known as *barytocelestite*. There are some rare varieties rich in Pb and Ra (hokutolite). Occasionally certain impurities such as Fe_2O_3, clayey, organic, and other matter.

Space group *Pnma(D$).*$a_0$ = 8.85; 6_0 = 5.43; c_0 = 7.13. Crystal structure. The Ba^{2+} and S^6+ ions are spaced at a distance of i and | of the *b* height from each other. The SO_4 groups are somewhat irregular tetrahedra which, are differently orientated. Each Ba ion is surrounded by twelve oxygen ions which belong to seven different SO_4 groups. Frequent barite crystals have generally a tabular habit owing to the development of the face. Less frequent are prismatic, crystals bounded with dominant prism faces in combination with isometric crystals. Occur in a great variety of combinations. Twins are rare; those observed are usually polysynthetic, producing striations on faces. Aggregates are commonly granular, less often compact, crypto-crystalline, and earthy. Also observed as stalactites and other flowstone sinter forms with a concentrical banded structure, and globular and ellipsoidal concentrations of radial structure. In cavities, often forms beautiful druses of small crystals.

Colour, Crystals may be colourless and water-transparent. However, commonly barite is coloured white or grey by foreign impurities (microscopic occlusions of gases and liquids), red (iron oxide), yellow or brown (probably iron hydroxides), dark grey and black (bituminous substances); sometimes it has bluish, greenish, and other tinges. Lustre, vitreous, pearly on cleavage surfaces.

Hardness, 3 to 3.5. Brittle. *Cleavage,* {001}, indistinct. *Specific gravity,* 4.3 to 4.5.

Diagnostic Features: Barite has the highest specific gravity among the most common sulphates (only the specific gravity of anglesite is higher). Characteristic features are also perfect unidirectional cleavage, insolubility in hydrochloric acid even on heating (in contrast to all the carbonates). Differs from certain similar silicates in cleavage and other features, and by much lower hardness. Hard to distinguish from celestite without chemical tests.

Decrepitates in the blowpipe flame, only the thin edge fragments become rounded. Colours the flame yellow-green (barium). With soda on platinum plate fuses into a transparent mass which on cooling becomes turbid (when fused on charcoal, the mass deliquesces and soaks into the coal). As different from celestite, barium sulphide formed in the reducing flame, when moistened with hydrochloric acid, colours the flame yellow-green instead of carmine-red. In the powdered state slowly dissolves in concentrated sulphuric acid. When water is added, the solution becomes turbid, i.e., $BaSO_4$ separates once more.

Genesis and Occurrence: In nature barite is formed in different ways, but only under conditions of increased partial oxygen pressure and at relatively low temperatures. Therefore, like all the other anhydrous sulphates, it never occurs as a magmatogenic product in igneous or deep-seated metamorphosed rocks.

In *hydrothermal* deposits it is rather common. As an accessory mineral, it is found in many sulphide, manganese (with manganite and braunite), iron (with siderite and haematite), and other ores. There are known gold-barite, almost purely barite, barite-calcite, barite-fluorite veins with a minor admixture of quartz and rare sulphides (galena, sphalerite, chalcopyrite, sometimes cinnabar, etc.).

In small amounts, mostly as concretions, barite is also widespread in *sedimentary* rocks, although in a different environment from that of anhydrite, gypsum, and celestite. Thus, it never occurs in salt deposits, very rarely in limestones, but often in sedimentary deposits of manganese (in oxide and carbonate ores) and iron, in argillaceous sandy, and other sedimentary deposits in near-shore zones of seas.

This is explained by the fact that the soluble salts of barium transported from land by surface water form a practically insoluble sulphate of barium at their very first encounter with $[SO_4]^2$ - ions in sea water. Nodular barite is found in silts also in contemporary seas.

In the zones of *weathering* of rocks and ore deposits in dry climate, close examination may reveal small barite crystals often of columnar habit associated with gypsum and iron hydroxides.

Barite is a chemically inert mineral and is, therefore, found in the eluvium often as coarse clastic material as well as in the heavy sands residue of placer sluicing operations. However, like all minerals of distinct cleavage and low hardness, barite soon becomes attrited in the places and gradually disappears.

The USSR has numerous barite deposits, of which we shall mention only the principal ones. Thus, in Western Georgia there are a number of vein deposits of barite at *Kutaissi, Bolnissi,* and other districts in tuffs and effusive rocks (porphyrites). Barite masses of symmetrically banded or colloform texture contain the following impurities:

Calcite, a negligible amount of quartz,

occasionally pyrite, chalcopyrite, and galena. The deposits of *Karakaly* district in the Kopetdag Ridge (Turkmenia) comprise a whole series of veins in sedimentary rocks (sandstones, clay shales, etc.). In some of the veins (Arpaklen) barite is associated with witherite, $BaGO_3$, which in places is pseudomorphous after barite.

A major foreign locality is *Meggen*, Westphalia (Germany), where a sheet-like body of extraordinary length (7 km) accompanied in places by a deposit of sulphides, mostly pyrite, lies at the contact of Mid-Devonian and Upper-Devonian sediments. The genesis of this particular deposit still remains obscure.

Uses: Barite is widely used in many industries.

1. In the form of fine powder as a weighting agent in drilling fluids for cementing unconsolidated rocks in oil-well drilling, for preventing blowouts and strengthening the walls of wells.
2. In chemical industry it is a source of various salts and agents for fireworks, treatment of hides (depilation) sugar refining, manufacture of photographic paper; in ceramics for making enamels, special grades of glass with high refractive indices, in medicine, etc.
3. In the manufacture of rubber and paper, as a filler and weighting agent.
4. For making high-quality lithopone (mixed with ZnO and ZnS) and paints, etc.
5. As the chief constituent of plaster far X-ray laboratories to protect the staff from harmful radiations.
6. Metallic barium is used in certain types of electronic tubes.

Celestite, $SrSO_4$: "Celestis" is the Latin for heavenly (the first discovered specimens were sky-blue in colour). Although relatively rare, it is the principal mineral of strontium.

Chemical Composition: SrO 56.4 per cent, SO_3 43.6 per cent. At times contains Ca and Ba (often in appreciable amounts).

System, orthorhombic; symmetry, rhombic dipyramidal $3L^23PC$. Space group *Pnma (D$)*. a_0= 8.36; b_0 = 5.36; c_0 = 6.84.

Crystal structure is exactly the same as that of barite. Crystals resemble closely those of barite. Usually tabular, columnar or prismatic. Aggregates mostly granular, sometimes rod-like, veined, and conchoidal, all of flowstone form. Also occurs in nodules and secretions and in druses in cavities.

Colour, Colourless white, often faintly blue or bluish grey, red or yellow. Occasionally colourless, water-transparent crystals are found. Lustre, vitreous to pearly on cleavage surfaces.

Diagnostic Features: In granular masses differs from carbonates of Mg, Ca, Sr, Ba, etc., by behaviour in acids (when dissolved with hearting, carbonates evolve CO_2). Sometimes resembles anhydrite in colour, but differs from it in the direction of cleavage and in higher specific gravity. Often hard to distinguish from barite. Very characteristic reaction for strontium.

Before the blowpipe fuses into a white head, colouring the flame intense carmine-red, especially when moistened with hydrochloric acid (strontium reaction). On charcoal with sodium carbonate gives characteristic sulphur liver. Soluble in strong sulphuric acid. The solution becomes turbid on addition of water.

Genesis and Occurrence: More or less considerable quantities of celestite occur as nodules, pockets, and massive bodies in *sedimentary* rocks (dolomites, limestones, gypsiferous clays, marls, etc.). Finds of celestite are often confined to definite horizons of the rocks.

Also found in marine organisms, viz., in skeletons of a certain group of radiolaria. Sometimes in the shells of ammonites and in other fossils as a result of later deposition.

Also occurs in typical but very rare hydrothermal veins

with galena, sphalerite, and other sulphides. Has been observed in amygdules of igneous rocks.

Celestite occurs fairly often as secretions in sedimentary gypsiferous Permian rocks in the *Archangel* Region, and Upper and Lower *Volga country* in *Bashkiria,* the *Orenburg Region,* etc. Numerous occurrences are known in the area of the *Caspian Sea* (in Turkmenia, Mangyshlak Island, etc.), and elsewhere in the Soviet Union.

Notable foreign localities are near *Bristol* (Britain), *Westphalia* and *Waldek* (Germany).

Uses: Gelestite is the main source of strontium salts used for fireworks (giving a bright carmine-red flame) and in chemical industry: sugar refining (as strontium oxide to extract sugar from molasses), in the manufacture of glass and ceramics (for iridescent glass, specially glazed bricks, etc.)- Lately, metallic strontium is used in special alloys, e.g., is added to copper to increase its strength and homogeneity (without decreasing its electrical conductivity).

Anglesite, $PbSO_4$

Discovered on the Anglesey Island (Wales), from which it derives its name.

Chemical Composition: PbO 73.6 per cent (Pb 68.3 per cent), SO_3 26.4 per cent. Usually free from impurities. A variety rich in BaO (8.45 per cent) is known.

Crystal structure exactly the same as that of barite. Very many combinations of crystal faces. Usually tabular, less often short-columnar, and pyramidal. Large crystals are rare. Aggregates. Often as crystalline crusts on galena, druses of small crystals, and in compact granular or earthy masses.

Colour, Colourless and often water-transparent. Often coloured grey, yellowish or brown by iron hydroxides, occasionally black owing to microscopic inclusions of undecomposed galena. Lustre, adamantine.

Hardness, 2.5 to 3. Very brittle. Cleavage, parallel to {001} distinct, parallel to {210} and {010} indistinct. Specific gravity, 6.1 to 6.4.

Diagnostic Features: Anglesite is recognised by high specific gravity, adamantine lustre, close association with galena in oxidised ores, and behaviour in the blowpipe flame.

In the blowpipe flame decrepitates and fuses readily. On charcoal with sodium carbonate gives sulphur liver and on further ignition in the reducing flame a lead globule.

Dissolves in concentrated sulphuric acid only on heating. Dissolves completely, however, in KOH (which distinguishes it from celestite and barite).

Genesis and Occurrence: As a poorly soluble product of the oxidation of galena and other sulphurous compounds of lead, anglesite is mostly formed in the *oxidised zones* of lead-zinc sulphide deposits, very often associated with the much more common cerussite; $PbCO_3$.

Microscopic examination shows that anglesite is the first oxygen compound of lead resulting from the oxidation of galena on the periphery and parallel to the cleavage, according to the reaction $PbS + 0_4 = = PbSO_4$. Under the action of carbon dioxide, however, anglesite readily changes in turn to lead carbonate, i.e., cerussite, $PbCO_3$. Being poorly soluble in water, both minerals usually form white crusts over massive galena which is thus protected from further oxidation. Therefore, when white nodules of seemingly completely oxidised lead ores are broken, their central portions often constitute solid galena.

It is also characteristic that the total lead content in the oxidised zones is invariably higher than in the primary sulphide ores. This is due to the fact that the constant accessory of galena—sphalerite—and copper sulphides are readily oxidised to well-soluble sulphates which are transported downward, thus enriching the oxidised ores with lead.

Anglesite may also be formed *hydrothermally,* though under specific conditions. Well-developed, often large, crystals occur in those lead-zinc ores which are deposited, judging by the paragenesis and the forms of mineral aggregates, near the daylight surface (probably through mixing with vadose waters saturated with free oxygen). Such are the deposits in Garinthia (the Western Alps) where anglesite crystals are found in cavities side by side with unoxidised sulphides of Fe, Zn, and Pb.

A certain amount of anglesite is always present in the oxidised zones of all galena-bearing sulphide deposits. Without enumerating all of them we shall merely mention that in Soviet literature there are described well-developed anglesite crystals from the Berezovskoye gold deposits (the Urals), from a number of deposits in the Altai mountains, Nerchinsk (eastern Trans-Baikal area), and elsewhere.

Uses: Anglesite mined together with other oxygen compounds of lead in the oxidised zones of lead deposits goes into the smelter.

Anhydrite and Gypsum

These two minerals stand apart both in physical and chemical properties from anhydrous and hydrous sulphates, but are closely linked with each other in the mode of formation and occurrence. Therefore, they will be discussed together.

The so-called calcium hemihydrate, which is easily obtained artificially, has not yet been positively recognised in nature.

Anhydrite, ($CaSO_4$)

The name of the mineral ("waterless") implies that in contrast to gypsum it contains no water.

Chemical Composition: CaO 41.2 per cent, SO_3 58.8 per cent. Quite frequently contains (010) plane 237, left) and on the (100) plane (closer to the observer) are different. The (010) face is centred, while on the (100) face, the Ca^2+ and $[SO_4]^{2-}$ ions do not form horizontal rows. Habit, thick-tabular or prismatic 239. Well-formed crystals are rare. Aggregates. Usually granular massive sometimes in rod-shaped aggregates.

Colour, white, often with a blue, greyish, sometimes reddish tinge. Colourless water-transparent crystals occur. Lustre, vitreous with pearly chatoyancy on cleavage (010). This becomes especially pronounced on heating. $n_x = 1.614$, $n_y = 1.576$, and $n_z = 1.571$.

Hardness, 3 to 3.5. Cleavage, {010} perfect, {100} and {001} distinct. In these three mutually perpendicular directions, crystals are rather easily split into cubic fragments. Specific gravity, 2.8 to 3.0 (for transparent varieties, 2.96). Other properties. In the presence of water under atmospheric pressure gradually changes to gypsum, greatly increasing in volume (up to 30 per cent). Increasing external pressure inhibits this conversion.

Diagnostic Features: Differs from the other sulphates of the group by lowest specific gravity; direction of cleavage planes; and optical properties (especially birefringence). Anhydrite also differs from marbled masses of carbonate (calcite, dolomite, and magnesite) in that it does not evolve CO_2 in acids. An hytrite is distinguished from gypsum by hardness (cannot be scratched with the fingernail).

Before the blowpipe fuses into a white enamel and colours the flame reddish-yellow. Does not fuse with soda or deliquesce and soak into the charcoal like barite, but decomposes and yields sulphur liver which stains a silver plate. In powdered form dissolves in sulphuric acid. The solution, in contrast to anhydrous sulphates of Ba, Sr, and Pb does not become turbid on addition of a moderate amount of water. Poorly soluble in hydrochloric acid.

Genesis and Occurrence: Anhydrite occurs in huge masses in thick sedimentary strata. As a product of chemical sediments (in lagoons and seas), anhydrite is almost invariably accompanied by gypsum. It is rather easily converted to gypsum in the outcrops. This conversion, according to the data obtained in numerous boreholes and mine workings, may occur down to a depth of 100 to 150 m from the surface (beyond which anhydrite masses are unaltered). At great depths the pressure of the overlying rocks is evidently so great that it prevents any increase of the rock mass volume which accompanied the alteration of anhydrite into gypsum.

Anhydrite often occurs in salt deposits either as individual crystals or as beds and intercalations (sometimes paper-thin) interbedded with halite, sylvite, carnallite, etc.

Relatively seldom anhydrite may be recorded from *hydrothermal* veins, and occasionally in contact-metasomatic deposits. Anhydrite had also occurred in voids in lavas in volcanic regions.

It is quite probable that the thick anhydrite strata in gypsiferous regions have derived from the dehydration of original gypsum strata caused by the great pressure of the overlying rocks. No anhydrite or any other sulphates are found, however, in deeper metamorphic rocks.

Great anhydrite deposits are found at depth in thick gypsiferous Permian strata all along the western foothills of the Urals, in the Archangel, Vologda, Kuibyshev, Gorky regions, Donets Basin, and elsewhere.

Uses. Anhydrite, as well as gypsum, is used chiefly in the production of binding materials (cements). Dense cryptocrystalline varieties are also used for ornamental purposes.

Gypsum, ($CaSO_4$-$2H_2O$)

Gypsum is the old Greek name for the mineral.

Chemical Composition: CaO 32.5 per cent, SO_3 46.6 per cent, H_2O 20.9 per cent. Usually pure. Mechanical impurities are clayey matter, organic substances (fetid gypsum), minute sand grains, sometimes sulphides, etc.

System, monoclinic; *symmetry*, prismatic L^2PC. Space group *A2ln(C&)*. a_0 = 10.47; 6_0 = 15.12; c_0 = 6.28; *ft* = 98°58'. Crystal structure. According to X-ray data, the mineral has a distinctly laminated structure. Two sheets of the $[SO_4]^{2-}$ anionic groups closely bonded with the Ca^{++} compose double layers oriented parallel to the (010) plane. These layers are arranged perpendicular to the plane of drawing and follow a vertical direction (the borderlines are dotted). The H_2O molecules are located between the double layers. This explains the characteristic perfect cleavage of gypsum. Each calcium ion is surrounded by six oxygen ions of the SO_4 groups, and by two water molecules. Each water molecule links the Ca ion with one oxygen ion in the same double layer and with another oxygen ion in the adjacent layer.

Habit: Owing to the predominant development of the {010} faces, crystals are tabular, rarely columnar or prismatic. The most common prisms are {110} and {HI}, sometimes {120}, etc. The faces {110} and {010} are often vertically striated. Penetration twins are common and may be of three types: (1) Gallio twins (100); (2) Paris twins (101), and (3) twins (209). It is not always easy to tell them from one another. The first two types have a swallow-tail shape. In Gallio twins the edges of the prism *m* {110} are parallel to the twinning plane, and the edges of the prism / {111} form a re-entrant angle, whereas in Paris twins the edges of the prism *I* {111} are parallel to the twinning plane 244). Aggregates. Occurs in cavities as crystal druses. Compact finely-crystalline aggregates are common. In cracks sometimes occurs in asbestos-like parallel-florous masses of gypsum with a silky chatoyancy and with the fibres perpendicular to the walls of the cracks. Such gypsum is known as selenite. When gypsum crystallises among loose sandy masses, it contains numerous entrapped sand particles which are clearly visible on the cleavage surfaces of large crystal individuals.

Colour, white, Individual crystals may be colourless and water-transparent. Also grey, honey-yellow, red, brown, and black (depending on the colour of the impurities entrapped during crystallisation).

Lustre, vitreous, with a pearly chatoyancy on cleavage surfaces. n_x = 1.530; n_y = 1.528, and n_z = 1.520.

Hardness, 1.5 (can be scratched with the fingernail). Very brittle. Cleavage, {010} excellent, {100} and {011} distinct. Cleavage fragments are rhombic with angles of 66° and 114°. Specific gravity, 2.3. Other properties. Well-soluble in water. A remarkable characteristic of gypsum is that its solubility is highest at 37°C to 38°C and then drops abruptly. The sharpest decrease in solubility begins at temperatures over 107°C, owing to the formation of the "hemi-hydrate" $CaSO_4.V_2H_2O$.

When heated under atmospheric pressure, gypsum, as evident from thermographs, starts losing water at 80° to 90° and at 120° to 140° completely changes to the hemi-hydrate, the so-called statuary or plaster gypsum known as plaster of Paris. On mixing this hemi-hydrate with water, a mortar is obtained which soon sets, expanding and evolving heat.

Diagnostic Features: A highly characteristic feature of crystalline gypsum is its perfect {010} cleavage and low hardness (can be scratched with the fingernail). Compact marble-like aggregates and fibrous masses are also identifiable by low hardness and the fact that they do not effervesce when moistened with hydrochloric acid.

Before the blowpipe loses water, decrepitates, and fuses into white enamel. In the reducing flame on charcoal gives GaS. Dissolves in water with a little sulphuric acid much more readily than in pure water. When the acid concentration exceeds 75 g/1, solubility drops sharply. Almost insoluble in hydrochloric acid.

Genesis and Occurrence: Gypsum forms in diverse ways in nature.

1. Large masses are formed by *sedimentation* in drying saline lacustrine and marine basins, where gypsum along with NaCl is formed only at the first stages of evaporation when the concentration of other salts in solution is not yet great. When such salts as NaCl and especially MgO_2 reach a certain point of concentration, anhydrite begins to crystallise, followed by other more soluble salts. Thus, gypsum in these basins must be one of the early chemical sediments. Indeed, in many

salt deposits the gypsum (and anhydrite) beds alternating with those of rock salt occupy the lower part of the deposit and are often underlain only by chemically sedimented limestones.

2. Considerable amounts of gypsum derive from the hydration of anhydrate in sedimentary deposits under the action of surface water under low external pressure (at the maximum average depth of 100 to 150 m), according to the following reaction:

$$CaSO_4 + 2H_2O = CaSO_4 . 2H_2O$$

The reaction is accompanied by a large increase in volume (up to 30 per cent) which involves numerous and complex local disturbances in the mode of occurrence of the gypsiferous strata. The majority of huge gypsum deposits in the world have been formed in this way. Massive gypsum in cavities sometimes contains pockets of large, often transparent crystals.

3. In deserts and semi-deserts gypsum frequently occurs as veinlets and nodules in the crust of weathering of rocks of most diverse composition. It may be formed on limestones by the action of water enriched in sulphuric acid or dissolved sulphates. Sometimes occurs in the oxidised zones or sulphide deposits, though not in as large amounts as might be expected. This is so because, in most cases, sulphide ores contain pyrite or pyrrhotite, the oxidation of which (especially of the former) substantially increases The sulphuricacid content of the surface water. The acidulated surface water considerably increases the solubility of gypsum. Hence, in a number of deposits gypsum is more common in the upper zones of primary ores, where it occurs in cracks together with other sulphates.

4. Gypsum is comparatively rarely found as a typically *hydrothermal* mineral in sulphide deposits formed at low pressures and temperatures. In such deposits it occurs sometimes in large crystals in cavities, and

contains inclusions of chalcopyrite, pyrite, sphalerite, and other minerals.

Such minerals as calcite, aragonite, malachite, quartz and other minerals frequently occur pseudomorphous after gypsum and vice versa.

Sedimentary gypsum deposits are found all over the world in strata of different geological ages. Without enumerating them, we shall merely say that in the USSR thick gypsiferous Permian strata occur in the Western Urals, Bashkiria and Tataria as well as in the Archangel, Vologda, Gorky, and other regions. Numerous Upper Jurassic deposits are found in the Northern Caucasus, Daghestan, Turkmenia, Tajikistan, Uzbekistan, etc.

Uses of gypsum are many and varied, especially in building:

1. Plaster of Paris (semi-calcined gypsum) is used for casts, moulds, "rock" lath, stucco, and wall plaster, in surgery, as filler for heavy white paper, etc.); in building as a binding material in masonry and for plaster floorings; in making bricks, window-sills, staircases, etc.
2. Uncalcined (natural) gypsum is used chiefly as a retarder in cement, for decorative and statuary purposes, as an ornamental stone (especially Ural selenite), in the preparation of paints, enamels, glazes, as a flux in smelting oxidised nickel ores, etc.

Sulphates of Alkali Metals

Besides anhydrous and hydrous sodium sulphates (thenardite and mirabilite) we shall describe in this group binary salts with alkaline earth metals (polyhalite and kainite).

Thenardite, (Na_2SO_4)

Chemical Composition: Na_2O 43.7 per cent, SO_3 56.3 per cent. Occasionally contains small amounts of K_2O, and as a mechanical impurity, $GaSO_4$.

System, orthorhombic, *symmetry,* rhombic dipyramidal Crystals are dipyramidal in habit, sometimes tabular, owing to the development of the {001} pinacoid and the *{101} prism*. Twins are frequent. Often occurs in druses or granular aggregates.

Colour, Colourless, transparent, sometimes with a reddish tinge. *Lustre,* vitreous. n_x = 1.485, n_y = 1.474, and n_z = 1.464.

Hardness, 2 to 3. Brittle. *Cleavage,* {010} perfect, {101} distinct, and {100} imperfect. *Specific gravity,* 2.66. Other properties. Soluble in water. Precipitates from oversaturated solutions (in pure water) only at temperatures over 32.5°C. Below this temperature crystallisation of mirabilite, $Na_2SO_4.10H_2O$, takes place. Therefore, at room temperature in humid air, thenardite is hydrated and covered with a white crust of hydrous sodium sulphate. With time decomposes to powder.

Diagnostic Features: In crystals differs from its mineral associates (mirabilite, astrakhanite, etc.) by crystallisation system, and in granular aggregates by chemical reactions.

Infusible in the blowpipe flame, colours the flame intense yellow (Na reaction). Readily soluble in water. Taste subsaline.

Genesis and Occurrence: Thenardite is formed in certain drying salt lakes together with mirabilite, crystallising directly from the brine supersaturated with Na and SO_4 ions. Also derives from the dehydration of mirabilite, e.g., in the upper layer of the material thrown up on the shore during storms (in the Caspian Sea), or in underground (i.e., not of recent formation) sodium sulphate deposits, and also in hot deserts. Thenardite dipyramidal crystal occurs in some volcanic regions (Vesuvius), at fumaroles.

A large buried deposit of thenardite in Pleocenic strata is located at *Kyuren-Dag* near the Azun-Su railway station, beyond the Caspian Sea. Together with mirabilite it occurs as lenses in sandy-clayey sediments. Also occurs in contemporary lakes in the *Kulunda Steppe* (North-Western Kazakhstan), north of the Caspian Sea (Lake Sha-shinskoye) in the *Shemakha* district (Transcaucasia), and elsewhere.

Uses: Extracted together with mirabilite and used in the manufacture of glass, soda, etc. Mirabilite, $Na_2SO_4.10H_2O$. The old name: "sal mirabile Glau-beri" (the wonderful salt of Glauber). Synonym: Glauber salt.

Chemical Composition: Na_2O 19.3 per cent SO_3 24.8 per cent, H_2O 55.9 per cent. Mirabilite owes many of its properties to high content of crystallisation water.

System, monoclinic; *symmetry,* prismatic. Crystals are short-columnar in habit parallel to the *b* axis or to the *c* axis. Usually in massive granular aggregates and also in crusts and coatings.

Colour, Colourless and transparent, sometimes turbid with yellow, bluish, or greenish tinges. *Lustre,* vitreous. Refractive indices extremely low: n_x = 1.398, n_y = 1.396, and n_z = 1.394.

Hardness, 1.5 to 2. Very brittle. *Cleavage,* {100} perfect. Fracture in other directions conchoidal. *Specific gravity,* 1.48. In dry air loses gradually all its water, changes to thenardite, and turns into a white pulverulent substance. Taste cool, slightly bitter-saline.

Diagnostic Features: As a mineral rich in H_2O molecules bonded within the crystal structure, mirabilite has the lowest specific gravity and refractive indices of all the readily soluble hydrous sulphates and chlorides. Differs from soda ($Na_2CO_3.10H_2O$) in that it does not evolve carbon dioxide under the action of hydrochloric acid. At 32°C deliquesces or rather dissolves in its own crystallisation water.

Genesis and Occurrence: Is formed in large masses in salt lakes saturated with sodium and sulphate anions in the course of evaporation of water at temperatures below 33°C (provided the solution contains no other soluble salts) or with decreasing temperature in autumn and winter. If evaporation of water from a solution of the same composition takes place at temperatures higher than 33°C, anhydrous sodium sulphate called thenardite will crystallise. The numerous small dried-up lakes of the Northern Caspian region, North-Eastern Kazakhstan (Kulunda Steppe), Northern Caucasus (south of Stavropol), Transcaucasia, the Crimea, etc., contain large masses of mirabilite and thenardite, with layers of halite, sometimes gypsum, and other minerals. The most important foreign localities for the occurrence of mirabilite are the Great Salt Lake in Utah (U.S.A.) and lakes in Mexico and Argentine.

Uses: Mirabilite is used mainly for containing potash by the Leblanc process (fusion with limestone and coal), as well as in the production of glass, paints, etc. It is used in medicine as a purgative.

Polyhalite, $K_2MgCa_2[SO_4]_4$-$2H_2O$

System, triclinic; *symmetry,* pinacoidal. Crystals rare, commonly tabular. Aggregates, compact, fibrous, rod-like.

Colour, white with a grey or yellowish tinge, and also brick-red. *Lustre,* vitreous. $n_x = 1.567$; $n_y = 1.562$; and $n_z = 1.548$.

Hardness, 2.5 to 3. *Cleavage,* {101} perfect. *Specific gravity,* 2.72 to 2.78. Before the blowpipe readily loses water and fuses into an opaque bead. Water leaches out the salts of K and Mg leaving a residue corresponding to gypsum in composition.

Polyhalite is formed from brines rich in Mg, K, and Ca within a relatively broad range of temperatures from 0° to 80°C. Occurs in deposits in the Stassfurt district as a "polyhalite stage" between the anhydrite and kieserite strata. Sometimes it is intercalated with thin layers of halite. Also occurs in salt deposits at Ischle, Halistadt (Austria), in Texas, New Mexico (U.S.A.), and elsewhere.

As a potassium salt it is used for the manufacture of artificial fertilizers.

Kainite, $KMg[SO_4]Cl_;3H_2O$ or $MgSO_4.KCl$-$3H_aO$

System, mono-clinic; *symmetry,* prismatic. Usually in granular massive. Rare crystals are either tabular or prismatic in habit.

Colour, yellowish- or greyish-white, sometimes red. Lustre, vitreous. $n_x = 1.516$; $n_y = 1.505$; and $n_z = 1.494$. *Hardness,* 2. *Cleavage,* {001} perfect. *Specific gravity,* 2.1.

Fusible in the blowpipe flame. Well soluble in water. Taste, bitter saline. Non-hygroscopic (in contrast to carnallite).

Widely occurs in the salt deposits at Stassfurt associated with schoenite, K^Mg$[SO_4]_2.6H_2O$; kieserite, halite, etc. Considerable kainite is found at Kalush (Western Ukraine).

A material for artificial fertilizers and potassium salts.

Hydrous Sulphates of Divalent Metals

Here we shall review some sulphates of magnesium, iron, and copper, especially those rich in H_2O.

Epsomite, $MgSO_4.7H_2O$: Named after the mineral springs at Epsom, England, in which it occurs. Synonym: Epsom salt.

Chemical Composition: MgO 16.3 per cent, SO_3 32.5 per cent, H_2O 51.2 per cent. May contain Mn" and Fe" as isomorphous admixtures. Certain varieties rather rich in divalent iron (ferroepsomite) and also in nickel (nickel-epsomite) have been established.

System, orthorhombic; *symmetry,* rhombic tetrahedral. A rather unstable monoclinic modification in the form of pseudohexagonal plates can be obtained artificially. Crystals precipitating from brines are pseudotet-ragonal prismatic or acicular. Occurs in compact sinters and earthy aggregates.

Colour, white; sometimes colourless and transparent.

Hardness, 2 to 2.5. Very brittle. *Cleavage,* {010} highly epsomite perfect, and {101} distinct. " crystal. *Specific gravity,* 1.68. Taste, bitter, subsaline. In dry air gradually loses water and becomes turbid. Dehydration proceeds unevenly: at 100° epsomite loses five water molecules, at 132°C the sixth water molecule, and at 218° to 238° the last molecule. On losing one water molecule in dry air, changes to magnesium hexahydrous sulphate.

Diagnostic Features: Epsomite is very similar to hydrous sulphates of more complex composition, and can be distinguished from them only by chemical tests. Has a characteristic bitter saline taste. Gives off water upon heating. On ignition yields an infusible white mass. Well-soluble in water.

Genesis and Occurrence: Epsomite forms in the course of evaporation of the brine of magnesia-rich salt lakes and is one of the first of the highly hydrated and magnesium sulphates to precipitate. As the brine gets thicker, magnesium heptahydrous sulphate becomes unstable yielding place to magnesium hexahydrous sulphate known as hexa-hydrite, $MgSO_4.6H_2O$.

Occurs as efflorescences on the surface of rocks, as crusts, capillary or acicular crystals on the walls of caves in calcareous rocks, abandoned mine workings, and other cavities near the surface; evidently as a result of crystallisation of percolating magnesial-sulphate waters. Epsomite is formed most frequently as crystalline precipitates in magnesial sulphate salt lakes: *Elton, Jelon, Malinov* in the lower Volga country in certain Crimean lakes (Sasyk-Sivash, and others); Jaman-Klych, north of the Sea of Aral in Kazakhstan, etc.

Notable foreign lakes are in the USA, Mexico, Egypt, China, and elsewhere.

Uses: Epsomite, as the other magnesium sulphates, is used in textile, paper-making, sugar-refining, chemical, pharmaceutical, and other industries.

Hexahydbite, $MgSO_4.6H_2O$***:*** System, monoclinic; symmetry, prismatic. Synonym: sacchite. Occurs in thick tabular, spear-shaped crystals or in fibrous masses. Twins on (001) and (110).

Colour, white, sometimes with a light-green tinge. Lustre, pearly. n_x = 1.456, n_y = 1.453, and n_z = 1.426.

Hardness, 2. *Cleavage,* {100} perfect. *Specific gravity,* 1.75. Well-soluble in water.

Occurs in magnesial-sulphate salt lakes in the Crimea, Astrakhan Region, and elsewhere. Precipitates from evaporating brine after epso-mite.

Kiesebite, $MgSO_4$-H_2O: Chemical composition. **MgO** 29.0 per cent, SO_3 58.0 per cent, H_2O 13.0 per cent.

System, monoclinic; *symmetry;* prismatic. Rare crystals are dipyramidal.

Colour, Seldom colourless and transparent; commonly turbid, white with a yellowish tinge. *Lustre,* vitreous. n_x = 1.584, n_y = 1.533, and n_z = 1.520.

Hardness, 3.5. Brittle. *Cleavage,* on {110} and {111} prisms perfect. *Specific gravity,* 2.57. In humid air gradually changes to epsomite. A powdered mass mixed with a small amount of water sets like plaster of Paris.

Diagnostic Features: Differs from similar hydrous sulphates (epsomite, kainite, etc.) in optical constants and composition. In the blowpipe flame decrepitates, loses water.and fuses readily. Slowly dissolves in water. Crystallises among the last minerals from complex, concentrated, magnesia-rich sulphate brines at temperatures above 18-20°.

Genesis and Occurrence: In salt lakes kieserite very rarely crystallises directly from the magnesia-rich sulphate brine (its range of stability at lower temperatures is very small).

Usually occurs in buried salt deposits probably deriving from the more water-abundant magnesium sulphates (epsomite and hexahyd-rite), through dehydration under pressure.

In quantity (up to 30 per cent of total mass), kieserite is found in the Stassfurt salt deposits (Germany) in association with halite, carnallite, polyhalite, anhydrite, etc.

Uses: A source of bitter salt (epsomite) the uses of which have been described above.

Melanterite, $FeSO_4 7H_2O$: "Melanteros" is the Greek for blacker. Probably, the first to be discovered was the rare grey-black variety. Is readily converted to black leather mordants and dyes.

Chemical Composition: FeO 25.9 per cent, SO_3 28.8 per cent,

H_2O 45.3 per cent. May contain isomorphous admixtures of Mg, Ni, Zn, Cu, and sometimes Mn.

System, monoclinic; *symmetry,* prismatic. *Habit,* rhombohedral, sometimes acicular or capillary. More often massive in the form of veinlets or sinters.

Colour, light-green, sometimes dark-grey, or greyish black. *Lustre,* vitreous. $n_x = 1.486$; $n_v = 1.478$, and $n_z = 1.471$.

Hardness, 2. Very brittle. *Cleavage,* {001} perfect, {110} distinct. *Specific gravity,* 1.8 to 1.9.

Diagnostic Features: May be distinguished positively from other readily soluble sulphates only by chemical analysis. A characteristic feature is the presence of divalent iron.

On heating deliquesces, dissolving in its own crystallisation water. On losing water turns into a pulverulent mass of anhydrous sulphate which turns brown in the reducing flame on charcoal (owing to the oxidation of its divalent iron).

Genesis and Occurrence: Melanterite crystallises from supersaturated sulphate waters under oxygen deficiency conditions. Therefore is never found in the oxidised zones. Usually occurs below the oxidised zone in cracks and cavities in semi-decomposed pyrite-rich ores in association with gypsum and other sulphates. Also occurs in clays in coal deposits (the source material for it is probably pyrite, often present in them).

In the USSR melanterite occurs in Kazakhstan, Central Asia, and Siberia. Found in the *Blyava* pyrite deposit (Southern Urals) as veinlets and accumulations in cavities in the upper horizons of semi-decomposed pyrite ores. Some black varieties of melanterite have been found there.

Uses: Natural occurrences are rarely large. Artificial melanterite is used as a mordant and in making paints: Prussian blue and black and dark dyes for woollens and leather and also for the preparation of ink and chemicals. It should be noted that heptahydrosulphate of ferrous iron with certain tanning agents yields a stable black pigment.

Chalcanthite, $CuSO_4.5H_2O$

"Chalcos" is the Greek for copper, "anthe" flowers ("copper flowers"). Synonym: copper vitriol.

Chemical Composition: CuO 31.8 per cent, SO_3 32.1 per cent, H_2O 36.1 per cent. Impurities: Fe, sometimes Zn, Co, and Mg (magnochalcanthite).

System, triclinic; *symmetry,* pinacoidal. Crystals are rare, short-prismatic and tabular. Usually occurs massive with stalactitic radial-fibrous structure.

Colour, azure, deep-blue, sometimes with a greenish tinge. *Lustre,* vitreous. n_x = 1.543, n_y = 1.537, and n_z = 1.514.

Hardness, 2.5. Very brittle. *Cleavage,* {110} imperfect. Fracture, conchoidal. *Specific gravity,* 2.1 to 2.3.

Diagnostic Features: Identifiable by sky-blue colour and high solubility in water; however for more positive identification, chemical analysis should be resorted to.

On heating gradually loses water and changes first to trihydrous sulphate and then to monohydrous sulphate, becoming white and opaque. Readily soluble in water yielding a blue solution. An iron needle immersed in the solution, is coated with metallic copper.

Genesis and Occurrence: Chalcanthite forms in the oxidised zones of copper sulphide deposits in dry climate. As stalactites occurs on the walls of abandoned unventilated mines driven through the oxidised zones of copper deposits.

In the USSR, chalcanthite occurs in a number of copper deposits: Mednorudyanskoye (Nizhny Tagil), Turya (Northern Urals), Kedabek (Transcaucasia), etc. Large masses are extremely rare.

Uses: Artificial chalcanthite is used as a pesticide in vineyards, and in the chemical dyeing, and other industries.

The presence of this mineral in ores shows that mine waters are rich enough in dissolved copper sulphate to warrant recovery of metallic copper by precipitation with scrap iron. This is done by passing mine waters through tanks in which copper is deposited on iron shavings.

Alunite Group

We shall consider under this heading a characteristic group of basic binary sulphates of Al and Fe with univalent (K, Na, NH_4) and some divalent (Pb) metals.

It should be noted that compounds with large univalent cations (K, NH_4) and occasionally with Pb^{2+} are pseudocubical and crystal-use in the ditrigonal pyramidal symmetry (in which the a-angle of the rhombohedron is close to 90°), whereas compounds with smaller cations (Na and Ag) are lower in symmetry order and pseudohexagonal.

Alunite, $KAl_3[SO_4]2[OH]_6$

Synonym: alumstone. The colloidal variety rich in adsorbed water is known as loewigite.

Chemical Composition: K_2O 11.4 per cent Al_2O_3 37.0 per cent, SO_3 38.6 per cent, H_2O 13.0 per cent, very often half of K_2O is replaced by Na_2O (natroalunite). Loewigite may contain rare earths.

System, trigonal; *symmetry,* ditrigonal pyramidal. Rare small crystals are rhombohedral, pseudocubical or thick-tabular. Usually in fine-granular, earthy, sometimes fibrous masses.

Colour, white with greyish, yellowish, and reddish tinges. *Lustre,* vitreous, sometimes pearly on *cleavage* surfaces. n_x = 1.592. Alunite crystal. n_y = 1.572 (for natroalunite 1.585).

Hardness, 3.5 to 4. *Cleavage,* {0001} distinct. *Specific gravity,* 2.6 to 2.8.

Diagnostic Features: In finely-crystalline masses alunite may be easily confused with many white hydrous minerals. The best way to identify it is by chemical tests.

Before the blowpipe decrepitates without fusing. Loses

water only on ignition. With soda yields sulphur liver. When moistened with a solution of cobalt nitrate assumes a deep-blue coloration (aluminium reaction). Insoluble in water and hydrochloric acid. Dissolves with difficulty in concentrated sulphuric acid. Ignited alunite placed in water gives alum in the solution.

Genesis and Occurrence: Alunite forms as extensive impregnations in igneous rocks rich in alunite-forming alkaline feldspars under the action of low-temperature sulphurous *hydrothermal* solutions. This alteration of rocks is called "alunitisation". Is also found in some hydrothermal veins.

Rare alunite concretions occur in sands, clays, bauxites, probably of exogenous origin.

As irregular yellowish or pinkish inclusions alunite is found in heavily altered tuffs (clastic rock of igneous origin) at *Zaglik* (Azerbaijan). At *Zhuravlinskoye* (25 km up the Chusovaya River from the Chusovskaya railway station) alunite with hydrargillite and kaolin forms pockets and veinlets of exogenous origin in the disintegration zone of limestones.

Uses: Alunite-bearing rocks are a source of alum and aluminium sulphate.

Jarosite, $KFe_3[SO4]_2[OH]_6$

Named after the place of discovery, Barranco Jaroso (Spain). Synonym: utahite.

Chemical Composition: K_2O 9.4 per cent, Fe_2O_3 47.9 per cent, SO_3 31.9 per cent, H_2O 10.8 per cent. Quite frequently contains small amounts of sodium and selenium and also SiO_2, Al_2O_3, etc., as mechanical impurities.

System, trigonal; *symmetry,* ditrigonal pyramidal. Often massive granular or earthy, also in druses of small rhombohedral crystals in cavities.

Colour, ochreous-yellow, often with a definite brown tinge. *Streak,* yellow. *Lustre,* vitreous to adamantine. n_y = 1.820, n_z = 1.715.

Hardness, 2.5 to 3.5. *Cleavage*, {0001} distinct. *Specific gravity*, 3.15 to 3.26.

Diagnostic Features: Externally resembles ochreous limonite. Differs from it by chemical reactions and also by greasy feel contrasting with the rough sand-like feel of limonite.

In the blowpipe reducing flame yields a magnetic mass. Does not dissolve in water, but is soluble in hydrochloric acid. Gives a $BaSO_4$ precipitate from a solution of BaCl. In the closed tube, gives water with an acidic reaction.

Genesis and Occurrence: Jarosite is a rather widely distributed mineral, formed in the *oxidised zones* of iron-sulphides, chiefly pyrite, deposits under the conditions of an arid climate and also after pyrite in certain sediments.

In moderate-temperature climates with abundant precipitation it is formed very rarely, since the iron sulphides decompose and are altered directly to iron hydroxides.

It has been demonstrated experimentally that jarosite precipitates from sulphate water contacting atmospheric oxygen and therefore containing trivalent iron. Jarosite is not very stable in water or air, yielding iron hydroxides on hydrolysis. Pseudomorphs of jarosite after cubic crystals of pyrite impregnated in rocks and after marcasite and pyrite concretions in clays are frequent.

Large masses of jarosite occur in the *Blyavinskoye* pyrites deposit (Southern Urals), in which it has survived in an old oxidised zone underlying an iron-hat zone composed of hydrohaematite and haematite; jarosite also occurs at *Maykain* (Bayanaul district, southwest of Pavlodar, Central Kazakhstan).

Uses: When found in large pure masses, jarosite is a material for polishing powders (Fe_2O_3) made by calcining.

ALUM GROUP

The term alum stands for normal binary sulphates of aluminium and alkalis crystallising in the cubic system with

12 water molecules. Halotrichite, i.e., sulphate of Fe and Al with 22 H_2O molecules, will also be described here.

Potassium Alum, $KAl[SO_4]_2$-$12H_2O$

System, cubic; *symmetry,* didodecahedral. Chemical composition. K_2O 9.9 per cent, Al_2O_3 10.8 per cent, SO_3 33.8 per cent, H_2O 45.5 per cent.

Colourless. Lustre, vitreous, n = 1.456 (often anisotropic). *Hardness,* 2. Specific gravity, 1.76. Readily fusible before the blowpipe. Soluble in water. Solubility at 20°C is 151 g/l.

Found in earthy masses, efflorescences, crusts, less often as massive granular aggregates.

Occurs in the so-called "alum earths" in the Tambov and Ulyanovsk regions, North Caucasus (Zelenchuk River), in Daghestan, Turkmenia (Kara-Kum desert), Uzbekistan (Shorsu), at the Blya-vinskoye pyrite deposit in the Urals (in the lower part of the oxidised zone together with other sulphates), on the right bank of the Oka River (a tributary of the Angara River in Siberia), etc.

Uses: Potassium alum has numerous uses: as a mordant in dyeing, in printing shops, in leather and paper manufacture, in medicine, for varnishes, etc. It is recovered by leaching from alum earths and subsequent evaporation of the solution. However, the bulk of alum is obtained artificially from bauxites or as a by-product of other chemical processes.

Soda Alum, $NaAl[SO_4]_2.12H_2O$

Synonym solfatarite. Has never been recognised positively in nature. Readily obtained artificially. Has been observed as parallel-fibrous masses in cracks in rocks.

Colourless, Lustre, vitreous. *Hardness,* 3. *Specific gravity,* 1.73. n == 1.439 (for the artificial compound). Natural specimens proved to be optically anisotropic, either because they were partially dehydrated or belonged to a different modification—mendozite. On exposure to air, soda alum is dehydrated and altered to tamarugite, $NaAl[SO_4]_2$ $6H_2O$. Has been described in the solfataras of Pozzuoli near Naples.

Halotrichite, FeAl2[SO4]4.22H2O

System, monoclinic; *symmetry,* dihedral axial L^2. Often occurs in asbestos-like veinlets of snow-white colour with a silky chatoyancy. Hence its name (translated into English means "hairy salt").

Hardness, 2 to 2.5. *Specific gravity,* 1.89 to 2.04. Dissolves in water. In the candle flame melts in its own crystallisation water and solidifies in drop-like forms.

It is formed from sulphate water rich in aluminium and ferrous oxide, under the conditions of acute deficiency of free oxygen below the oxidised zone of pyrites, as, for example, in the Blyavinskoye deposit (Southern Urals).

Also found in pyrite-bearing clays and in coal seams where it is formed as a product of the decomposition of pyrites in the oxidised zone and of the decomposition of argillaceous matter by the evolving sulphuric acid.

Chromates

Naturally-occurring salts of chromic acid are very few. The $[CrO]^{2-}$ complex anion is capable of forming crystal structures with the large Pb^{2+} cation, Chromate of K^{1+} are extremely rare.

Chromates are formed only in an environment with high oxygen concentration. Although they may occur with sulphates, chromates form neither isomorphous mixtures nor binary salts with them. It has been shown experimentally that $PbCrO_4$ and $PbSO_4$ can intermix in the solid state only at high temperatures. It also has been shown that the metastable $PbCrO_4$ modification crystallising in the ortho-rhombic system is isostructural with $BaSO_4$.

Crocoite, $PbCrO_4$

Crocos" is the Greek for saffron (named for its orange-yellow colour). The element Cr was discovered in crocoite from the Urals in 1797. The mineral has first been recognised in the days of Lomonosov (in the middle of 18th century).

Chemical Composition: PbO 68.9 per cent, CrO_3 31.1 per cent. May contain a little silver. The zinc-bearing variety called iossaite has a specific gravity of 5.2 (has not been studied in detail).

System, monoclinic; *symmetry,* prismatic. Crystals are frequent and occur exclusively in cracks. *Habit,* prismatic parallel to the *c*-axis, sometimes acute rhombohedroidal with {110} and {101} faces and octahedroidal {111} faces. Also in druses of rod-like crystals in cracks.

Hardness, 2.5 to 3. Brittle. *Cleavage,* {110} distinct. *Specific gravity,* 6.0.

Diagnostic Features: Highly characteristic features are bright red colour, rod-like or prismatic crystal habit and high specific gravity.

Readily fusible in the blowpipe flame giving a lead globule. The bead of salt of phosphorus. Crocoite and borax assumes an emerald-green colour crystals. (chromium reaction). Soluble in hot hydrochloric acid with evolution of chlorine and precipitation of $PbCl_2$. Turns brown and dissolves in KOH.

Genesis and Occurrence: Crocoite is formed in the oxidised zone of lead ore deposits near ultrabasic igneous rocks which, upon weathering, may yield at least minor amounts of chromic acid evolving through the disintegration of chromium-bearing silicates and chrome-spinellids. Very well known is the crocoite from the *Berezovskoye* gold deposit (near Sverdlovsk), first described in 1766 by Academician I. G. Leman. It occurs in cracks in beresite (altered granite-porphyry) as elongated prismatic crystals oriented parallel to the walls. Sometimes associated with galena. On *the Island of Tasmania* large druses of crocoite crystals up to 10 cm long occur in serpentinites. Also occurs in many other deposits.

Because of its sparse distribution is of no commercial importance.

Molybdates and Tungstates

Molybdenum and tungsten are found in the fifth and sixth periods of the periodic system. These elements, just as other pairs in the periods, viz., Zr and Hf, Nb and Ta, are characterised by almost equal atomic and ionic radii owing to the so-called lanthanidic compression, and hence have equal atomic and ionic volumes. Therefore, both elements, as the other pairs, might be expected to yield wide series of isomorphous mixtures in the course of the formation of minerals.

In nature, however, isomorphous mixtures of compounds of this class are very rare. Molybdenum and tungsten play materially different parts in the geochemical processes of mineral formation. Molybdenum, as has been mentioned when sulphides were discussed, has a distinct affinity for sulphur. The bulk of this element occurs in the earth's crust as a *sulphide*, MoS_2. Spectral and chemical analyses show its presence as an admixture to many sulphurous compounds. Its oxygen compounds seldom occur in nature and are chiefly confined to the *oxidised zone* of ore deposits.

Tungsten, on the other hand, yields almost exclusively *oxygen* compounds in the form of tungstates in the processes of mineral formation. Tungsten sulphide known as tungstenite, WS_2, is so rare that it has never been found in a quantity sufficient for detailed mineralogical study. Only recently it has been found in a more considerable quantity in the form of pseudomorphs after scheelite that have been studied by O. M. Rimskaya-Korsakova. It should also be noted that most tungstates are relatively *high-temperature* formations.

Among the few minerals of this class two principal groups stand out.

1. Tungstates of Fe, Mu, and Zn. Molybdates of these metals are so far unknown, with the exception of ferrimolybdite.
2. Molybdates and tungstates of stronger divalent metallic ions with larger ionic radii, such as Ca, Pb, and partly Cu.

Wolframite Group

Included in this group are the minerals of the $MnWO_4$ — $FeWO_4$ (hiibnerite-ferberite) isomorphous series. Since the minerals have many common properties, we shall describe in detail only the most widespread wolframite, (Mn, Fe)WO_4, a species of intermediate composition, mentioning in passing the important characteristics of the end members of this series.

Wolframite, (Mn, Fe)WO4

At first it was named wolfram (1758) which is the German translation of "Lupi spuma", a Latin term meaning "wolf foam" or "wolf cream", deriving from the property of the mineral, when admixed to tin ores, to produce foam on the surface of the smelted tin. The element tungsten was discovered later; sometimes it is also called wolfram after the mineral.

Chemical Composition: A. K. Boldyrev, suggested to classify as wolframite the varieties containing 25 to 75 molecular per cent of $MnWO_4$, i.e., 5.9 to 17.6 weight per cent of Mn. Varieties containing less than 5.9 per cent Mn are called ferberite, and those containing 17.6 to 23.4 per cent, hiibnerite. The WO_3 content is about 75 per cent. Impurities are Mg (up to 0.5 per cent), sometimes CaO, Ta_2O_5, Nb_2O_5, SnO_2, etc. Wolf- System, monoclinic; symmetry, prismatic.

Frequent crystals are thick tabular or prismatic sometimes flattened parallel to {100}. Some individuals from certain deposits are very large (20 cm and larger). Vertical faces are often striated parallel to the *c*-axis. Twins are parallel to (100), sometimes to (001). Also occurs in massive coarse-granular aggregates.

Colour, brownish black. *Hiibnerite* usually has a pinkish or violet tinge; ferberite, black. Streak, brown, for ferberite darker to black, for hiibnerite lighter to yellowish brown and even yellow. Lustre, on cleavage surfaces mirror-like, adamantine and greasy in other directions. Hiibnerite in thin fragments is translucent. Refractive indices for hiibnerite are n_x = 2.32,

n_y = 2.22, and n_z = 2.17, becoming higher with increasing iron content: in the almost opaque ferberite n_y goes up to 2.40 (Li-light).

Hardness, 4.5 to 5.5. *Brittle. Cleavage,* {010} perfect. *Specific gravity,* 6.7 to 7.5. Ferruginous varieties have a relatively high specific gravity. Iron-rich varieties are weakly magnetic.

Diagnostic Features: Wolframite has characteristic brownish-black colour, dark-brown streak, high specific gravity, perfect unidirectional cleavage (in distinction from black sphalerite, whose cleavage is multidirectional, and also from columbite, tantalite, cassiterite, whose cleavage is imperfect).

On intensive ignition in the blowpipe flame on charcoal fuses into a magnetic globule. When a wolframite-soda flux is boiled with tin in water acidulated with hydrochloric acid, it gives a light-blue solution (W reaction). Gives the Mn and Fe reactions with borax. Insoluble in hydrochloric acid.

Genesis and Occurrence: Wolframite occurs chiefly in quartz and *hydrothermal quartz* veins often confined to granite masses. Occasionally associated with cassiterite, molybdenite, arsenopyrite, pyrite, chalcopyrite, etc.

In *greisens,* i.e., in the portions of granite masses heavily altered by pneumatolytic action, wolframite is associated with such minerals as micas, topaz, fluorite, tourmaline, sometimes with beryl, cassiterite, molybdenite, etc.

Sulphide veins containing wolframite in paragenesis with chalco-pyrite, molybdenite, pyrite, bismuthinite, sphalerite, etc., are known.

Less often and usually in small quantities, it occurs in *pegmatite* dikes in granites.

Mineralogically interesting is the frequent endogenous replacement of scheelite crystals by wolframite both on the periphery and along the cracks. In turn, scheelite often replaces wolframite.

In the oxidised zone exposed to weathering, wolframite is

altered, though with difficulty, into the so-called tungsten ochre. In this process, divalent iron is oxidised into the trivalent modification. As a result, the crystal lattice breaks down and earthy yellow-brown or brown masses are formed, consisting mainly of hydrotungstate of trivalent iron (ferritungstite). Sometimes yellow-green tungsten oxides called tungstite, or meymacite, H_2WO_4, are formed.

Hibnerite, which breaks down in much the same manner, yields black "psilomelanic" accumulations containing WO_3. It should be noted that the psilomelane nodules present in eluvium, even at a considerable distance from the primary deposit, contain up to a few per cent WO_3.

Commonly, however, in the vicinity of primary deposits, wolframite, as a relatively inert mineral, usually passes into placers. But it is characteristic that with increasing distance from the primary deposits, the fragments of the wolframite minerals are comparatively quickly attrited and finally disappear altogether. This is due to the relative brittleness of the minerals which is enhanced by their perfect cleavage.

The world's largest wolframite deposits are in South China, Yunan and Kiangsi provinces (Hsihuashan), and also in northern Viet-Nam, South Burma and the Malacca Peninsula. In the United States, there are a number of deposits in Colorado, South Dacota, Nevada, and elsewhere. In Europe, wolframite deposits are found in Portugal and Spain.

Uses: The minerals of this group are the principal source of tungsten which has numerous applications.

1. Non-ferrous metallurgy consumes 85 to 90 per cent of the total output of tungsten for special self-hardening steels used in high-speed cutting tools.
2. Tungsten is a component of the so-called stellites, i.e., alloys of tungsten with Cr, Co, etc., which are employed in special-purpose tools and superhard alloys such as "pobedit", "vidia", "volomite", etc. These alloys are used for the manufacture of drill bits.

3. Tungsten is used for incandescent lamp filaments, targets in X-ray tubes, etc.
4. Tungsten compounds are used in the chemical industry, in colouring glass and porcelain, and for other purposes.

Scheelite Group

This group comprises both tungstates and molybdates of Ca and Pb crystallising in the tetragonal system; only a lead tungstate modification crystallising in the monoclinic system is known. Powellite, $CaMoO_4$:

Chemical Composition: CaO 28 per cent, MoO_3 72 per cent. Some varieties contain an isomorphous admixture of WO_3 (up to 8 per cent).

System, tetragonal; *symmetry,* dipyramidal. Crystal structure, analogous to that of scheelite. Occurs as earthy-foliated pseudomorphs after molybdenite. Separate crystals are dipyramidal in habit, sometimes tabular along {001}. The faces {112} are often obliquely striated.

Colour, pale yellow, in crystals yellow-green. *Streak,* light with a yellowish or greenish tinge. Lustre, adamantine. n_x = 1.984 and n_y - 1.974.

Hardness, 3.5. Brittle. *Cleavage,* none. *Specific gravity,* 4.25 to 4.52.

Diagnostic Features: Powellite pseudomorphs after molybdenite are readily identifiable by relict platy forms and pale-yellow colour. Crystals are characteristically dipyramidal and give the Mo and Ca reactions.

Before the blowpipe fuses into a semi-transparent mass. With salt of phosphorus in the oxidising flame yields yellowish-green bead, and a dark-green bead in the reducing flame. Soluble in hydrochloric acid. A dilute HCl solution in the presence of excess NH_3 with ammonium oxalate gives a calcium oxalate precipitate.

Genesis and Occurrence: As has been stated earlier,

powellite often occurs as pseudomorphs after molybdenite in the *oxidised zones* of molybdenum deposits, deriving probably from the reaction of molybdenic acid, formed by oxidation, with calcium-bearing surface waters. With time it is gradually leached out, leaving cavities in quartz in place of the molybdenite crystals.

Exogenous powellite frequently occurs in virtually all oxidised zones of molybdenite deposits, whereas endogenous powellite is rare.

Uses: As compared with molybdenite, is of secondary importance, because it rarely occurs in quantity.

Scheelite, $CaWO_4$

Named after the Swedish chemist Scheele (1742-1786), the discoverer of tungstatic acid in this mineral (tungsten in wolframite was discovered later).

Chemical Composition: CaO 19.4 per cent, WO_3 80.6 per cent. Occasionally contains an isomorphous admixture of MoO_3 (up to 10 per cent). There is also a copper-bearing variety called *cuproscheelite* with a CuO content of up to 7 per cent. At times contains rare earths (mostly of the cerium group). System, tetragonal; symmetry, dipyramidal. Habit, octahedral, sometimes tabular along {001}. The faces (112) sometimes show oblique striations. Penetration twins on (110) and (001) are fairly often. Aggregates Mostly as irregular inclusions, less often massive.

Colour, Rarely colourless, usually grey, yellow, greenish yellow, brown.

Hardness, 4.5. Brittle. *Cleavage,* {111} distinct. Fracture uneven. *Specific gravity,* 5.8 to 6.2. Other properties. In cathode rays usually fluorescos bright blue.

Diagnostic Features: In silicate rocks scheelite is sometimes scarcely noticeable. Can be recognised by the naked eye only through long experience. In crystals it is identifiable by octahedral cleavage, while on fracture and in aggregates by rather strong slightly greasy lustre, rather distinct cleavage, high specific gravity, and reaction for tungsten. Highly characteristic is its blue fluorescence in cathode rays. This property is widely made use of for identification in situ and even for a rough estimate of its content in the ore.

Fusible with difficulty in the blowpipe flame. Soluble in hydrochloric and nitric acids, decomposes with evolution of yellow tungstatic acid, H_2WO_4 H_2O, which is soluble in ammonia. Solution in hydrochloric acid when boiled with tin acquires a beautiful deep-blue colour.

Genesis and Occurrence: Scheelite occurs relatively often as a *hydro-thermal* mineral in deposits of different ores.

Small amounts are sometimes found in pegmatites. The biggest deposits, however, are of *contact-metasomatic* origin. In these it is associated with silicates (garnets, pyroxenes, etc.), quartz, and frequently with sulphides, such as molybdenite. It is often found in wolframite-bearing, auriferous, and other vein deposits.

In the oxidised zone scheelite is not very stable. On the surface, quartz veins sometimes have cavities in the place of leached scheelite. Nevertheless, scheelite quite often shows in the heavy concentrates of placer washing.

Notable foreign localities for the occurrence of scheelite are the contact deposits in western USA, the great deposit in Malaya where it often occurs in association with cassiterite in metamorphosed limestones and schists, as well as in quartz veins, also in northern *Tasmania* and elsewhere.

Uses: Both scheelite and wolframite ores are a source of tungsten.

Wulfen1te, $PbMoO_4$

Chemical composition. PbO 61.4 per cent, MoO_8 38.6 per cent. Occasionally contains admixtures of CaO, CuO, MgO, WO_3, rarely Cr_2O_3 and V_2O_5.

System, tetragonal; symmetry, pyramidal. Habit often in

square tablets, less often in combinations of obtuse and acute pyramids. Massive crystal aggregates are rare.

Colour, is wax- or honey-yellow, grey, brown, sometimes orange, or even red. Streak, white or very faintly coloured. Lustre, adamantine, greasy on fracture. $n_y = 2.40$ and $n_z = 2.28$.

Hardness, 3. *Cleavage,* {111} distinct. *Specific gravity,* 6.3 to 7.0.

Diagnostic Features: Characteristic honey-yellow colour, tabular habit, adamantine lustre, high specific gravity, and paragenesis with other minerals of lead in the oxidised zone.

Fusible in the blowpipe flame. On charcoal with soda yields a lead globule. With salt of phosphorus in the oxidising flame gives yellow-green, and in the reducing flame, dark-green bead (Mo reaction). Dissolves slowly in hydrochloric acid becoming covered with a white coating of Pbo_2.

Genesis and Occurrence: As a salt poorly soluble in water it is rather often found in the *oxidised zones* of lead-zinc deposits. May derive its molybdenum from two sources: either from the molybdenic acid brought by waters percolating from the wall rocks or through oxidation and concentration of the molybdenum dispersed in the sulphides within the deposit.

Usually wulfenite occurs as small crystals and crystalline crusts on the walls of leaching cavities. It may also occur pseudomorphous after other lead minerals in the oxidised zone, in particular cerussite. In this case, wulfenite is associated with cerussite, anglesite, and galena. It is also found in low-temperature hydrothermal deposits of lead-zinc ores.

Uses: When found in quantity, wulfenite together with other oxidised lead minerals, is a source of lead and molybdenum.

Hydrous Molybdates and Hydrous Tungstates

This division includes a very brief description of as yet inadequately studied hydrous salts of trivalent iron.

Ferrimolybdite, Fe2 [MoO$_4$]$_3$.7H2O

System, orthorhombic. Occurs in fibrous or scaly masses of sulphur-yellow colour. n_y = 1.87 to 2.05, n_x = 1.753 to 1.79, and n_z = 1.72 to 1.78.

Hardness, 2. *Cleavage*, {001} distinct. *Specific gravity*, 4.5. Readily fusible. Soluble in acids and NH_4OH.

An alteration product of molybdenite. Occurs in the oxidised zones of ore deposits.

Ferritungstite, $Fe_3[WO][OH]_4.4H_2O$

System, hexagonal. Occurs in microscopic plates and scaly aggregates of light-yellow or brownish-yellow colour. n_y = 1.80, and n_z = 1.72. Other physical properties as yet unknown. Decomposes in acids with evolution of yellow WO_3. As a product of wolframite oxidation is found in the Deer Trail deposit, in Washington (U.S.A.), and elsewhere.

Phosphates, Arsenates, and Vanadates

This class includes comparatively numerous mineral species of diverse composition. Their total quantity by weight in the earth's crust is relatively small, however.

The $[PO_4]^{3-}$, $[AsO_4]^{3-}$, and $[VO_4]^{3-}$ trivalent anions are rather large, and therefore the most stable *anhydrous* compounds may be expected from the combination of these anions with tri-valent large cations. In the introduction to the description of oxygen salts, it was stated that such cations are ions of rare earths and Bi. Compounds with small cations are, as a rule, much more extensively represented among all typically oxygen salts by *hydrous* normal salts (with hydrated cations).

Phosphates of divalent metals, with additional anions (OH, F, O sometimes Cl) or in the form of acid phosphates are most stable, also when the compounds contain relatively large cations (Ca,Sr, and sometimes Pb); characteristic for arsenates and vanadates are compounds with Pb, Cl being an additional anion, and finally binary compounds of Pb or Ca with Cu, Zn,

Mg, and sometimes with Mn. Calcium is also known to form hydrous acid salts.

In this connection it would be relevant to compare dihydrous salts, the phosphate and arsenate of Ca, with the sulphate of Ca (gypsum), all of which crystallise in the monoclinic system and have many properties in common.

Gypsum, $CaSO_4 \cdot 2H_2O$......... 10.47, 15.15, 6.28, 98°58', 2, 2.31

Brushitc, $CaHPO_4$-$2H_2O$...... 10.47, 15.15, 6.28, 95°15', 2, 2.25

Pharmacolite, $CaHAsO_4.2H_2O$ 10.97, 15.40, 6.29, 96°36', 2-2.5, 2.64

This is another proof that the H^{1+} cation, being very small, has practically no effect on the size of the crystal structure. The $[SO_4]^{2-}$ and $[PO_4]^{3-}$ anions, as seen from the above, are of exactly the same size, and the $[AsO_4]^{3-}$ anion is somewhat larger. While the refractive indices in these minerals increase with anion size from gypsum to pharmacolite, the birefringence of all these minerals remains practically the same. The specific gravity of the minerals fully corresponds to the atomic weights of the cations in the complex anions (S^6+, P^5+, and As^5+).

As it should have been expected, the phosphates and arsenates of divalent small cations (Mg^{2+}, Fe^{3+},Ni^{3+}, CO^{3+}, Cu^{2+}", Zn^{2+}) form highly characteristic hydrous normal salts.

Univalent metals (Na, Li) form, as a rule, binary compounds with Al^{3+}, and also a series of rare complex hydrous salts.

A special place among hydrous salts, with regard to composition, is occupied by the so-called uranmicas and complex vanadates.

Highly interesting are the phenomena of heterovalent isomorphism which are quite frequent in this class of compounds, especially in phosphates. Most vivid examples are the isomorphous substitutions of anionic radicals.

Thus, the $[PO_4]^{3-}$ trivalent anion may be replaced by similarly built and equidimensional anions such as divalent $[SO_4]^{2-}$ and tetra-valent $[SiO_4]^{4-}$. This substitution may proceed in a variety of ways:

1. Isomorphous substitution in the *anionic portion of* a compound takes place without affecting the composition and charge of the cationic portion. In this case, the magnitude of the total anionic charge should remain unchanged. This happens when the divalent $[SO_4]^{2-}$anion enters into the composition of the mineral simultaneously with the replacement of the trivalent $[PO_4]^{3-}$ anion by the tetravalent $[SiO_4]$ - anion. Only in this case can the total anionic charge remain unchanged.

In this case, the crystal structure and physical properties of the minerals naturally remain the same. It may be added that in the cationic portion too the Ca^{2+} ions may be simultaneously replaced by equidimensional Na, Y, and TR ions (with the composition and charge of the anionic portion remaining unchanged).

2. Isomorphous substitution in the anionic portion of a compound may take place simultaneously with the replacement of ions in the cationic portion by cations of a different valency. Thus, it has long been known that monazite, $GePO_4$, occasionally contains considerable SiO_2, i.e., the $[SiO_4]^{4-}$ anion, and that at the same time a corresponding amount of Th^{4+}, and occasionally U^{4+} and Zr^{4+}, is isomorphously admixed to Ce^{3+}. Recently Soviet scientist I.D. Borneman-Starynkevich showed by means of delicate chemical analyses that in certain monazites the isomorphous admixture of Ca^{2+} is accompanied by the divalent $[SO_4]^{2-}$ anion entering into the composition of the anionic portion. Therefore, the general chemical formula for such monazite varieties should be written as (Ce, Th, Ca) $[PO_4, SiO_4, SO_4]$.

It should be noted that heterovalent isomorphism does not necessarily require a strict quantitative balance between ions of the cationic and anionic portions. However, it does require: (1) equality of positive or negative total charges; (2) equality or proximity of the size of replacing ions, and (3) preservation of the total number of cations and anions in the course of replacement (except when one of the cations is an H^{1+} proton).

We shall not dwell on the general characteristics of physical properties of the most diverse minerals of this class. Since they vary widely, it seems more appropriate to review them for each individual group of minerals.

As for the mode of formation of the vast majority of minerals of this extensive class, especially hydrous compounds, they are associated with exogenous processes (in the oxidised zones of ore and sedimentary deposits). The endogenous minerals are almost exclusively the phosphates the majority of which is formed during the final stages of magmatic processes, mainly in pegmatites, and sometimes in hydrothermal veins. Apatite (calcium phosphate) also occurs in magmatic deposits (in nepheline-syenites).

Since the salts of this class include widely diverse hydrous and anhydrous compounds, all these minerals, in distinction from sulphates, fall into two large groups:

(1) Anhydrous phosphates, arsenates and vanadates;

(2) Hydrous phosphates, arsenates and vanadates.

Anhydrous Phosphates, Arsenates, and Vanadates

The anhydrous compounds of this class are represented in nature by a considerable number of minerals. In addition to the *normal* salts of orthoacids there are some rare species of *acid* salts. Much more common are *basic* salts.

Monazite Group

Monazite, (Ce, La)PO_4.

"Monazen" is the Greek for solitary (hence the word monk). Usually occurs as single crystals.

Chemical Composition: The content of oxides of rare earths (chiefly Ce and La) may be as high as 50 to 68 per cent; may contain an isomorphous admixture of Y_2O_3 (up to 5 per cent).

P_2O_5 content usually from 22 to 31.5 per cent. Often contains also an isomorphous admixture of ThO_2 (5 to 10 per cent, sometimes as high as 28 per cent), in some instances ZrO_2 (up to 7 per cent) with SiO_2 (up to 6 per cent), and sometimes GaO accompanied by SO_3.

Thus, from the standpoint of crystallochemistry, the mineral provides a graphic example of heterovalent isomorphism. The chemical formula for such varieties of monazite should look as follows: (Ce, La, Th, Ga) [PO_4, SiO_4, SO_4]. Negligible amounts of MgO, MnO, PbO, Fe_2O_3, Al_2O_3, and H_2O are also present.

System, monoclinic; *symmetry*, prismatic. Habit, usually tabular {100} 255, less often prismatic, isometric, and pyramidal. Most common forms are pinacoids {100}, {010}, {101}, prisms {110}, {011}, etc. Often striated. Twins on (100) and (001). Usually occurs as small crystals. Some large individuals, however, have weighed up to several kilograms.

Colour, yellowish brown, brown, red, rarely green. Lustre, strong vitreous, greasy. n_x = 1.837, n_y = 1.787, n_z = 1.785.

Hardness, 5 to 5.5. *Cleavage*, sometimes perfect parallel to {001}. *Specific gravity*, 4.9 to 5.5. Other properties. Often radioactive (the presence to ThO_2).

Diagnostic Features: In pegmatites arid granites monazite crystal is usually identified by characteristic tabular habit of opaque crystals of dark-yellow or reddish-brown colour.

Almost infusible in the blowpipe flame. Powdered monazite moistened with sulphuric acid colours the flame green. Dissolves with difficulty in hydrochloric acid giving a white precipitate. With borax, gives a yellow or yellowish-red bead when hot, which becomes colourless on cooling.

Genesis and Occurrence: The mineral usually occurs in *pegmatites*, sometimes in *granites* and *gneisses* in paragenesis with feldspars, zircon ($ZrSiO_4$), in some cases, with magnetite, ilmenite, and other minerals. Occasionally found in dolomite veinlets of hydrothermal origin in association with magnetite.

Upon weathering of primary deposits, chemically inert monazite is concentrated in valley and beach placers.

Notable monazite deposits are in the pegmatites of *Madagascar* (Ancasoba and Mount Volhambohitra) where very large crystals are found. Also occurs in very large valley and beach placers in *Brazil* (Minas Geraes) where it is recovered on a commercial scale.

Uses: Monazite is the chief source of thorium and the rare earths and is mined almost exclusively from placers.

Rare earths, in the main cerium, have most diverse uses. At present they are employed chiefly in pyrophoric alloys and cerium carbon-arc electrodes for search lights. They are also used in the manufacture of special glass permeable to light but impermeable to ultraviolet and most of heat rays, in strong and light alloys (of cerium with magnesium, aluminium, etc.), as catalysts in the production of chemicals, in dyes, etc.

Xenotime, YPO_4: "Xenos" is the Greek for alien, "timi", honour.

Chemical Composition: The theoretical Y_2O_3 content should be 63.1 per cent. Commonly contains small amounts of rare earths (Er, Ce) sometimes ThO_2, UO_2 (up to 5 per cent), ZrO_2 (up to 3 per cent), SnO_2,SiO_2 (up to 9 per cent), etc. A variety known as *hussakite* contains up to 6 per cent of SO_3 and probably CaO.

System, tetragonal; *symmetry,* ditetragonal dipyramidal. Occurs in prismatic crystals similar to those of cassiterite and zircon, and also massive. Closely resembles the rutile group in crystal structure. Crystals are usually a combination of {HO}, {010} and {011}. Other forms are rare.

Colour, yellowish brown, red, grey. *Streak,* light brown or reddish. *Lustre,* vitreous, waxy, or greasy. n_x= 1.815 and n_y= 1.720.

Hardness, 4 to 5. Fracture uneven. Brittle. Cleavage, {100} perfect. *Specific gravity,* 4.45 to 4.59. Often radioactive.

Diagnostic Features: Easily distinguished from zircon, rutile, and cassiterite by lower hardness and, in thin polished sections, by different optical properties.

Infusible in the blowpipe flame. Insoluble in usual test acids. When moistened with sulphuric acid, colours the flame bluish green.

Genesis and Occurrence: Occurs in granites and pegmatites as small disseminated crystals, often in association with zircon, with which it may develop oriented intergrowths, with apatite, orthite, monazite, etc.

As a relatively inert compound, xenotime passes into placers upon the destruction of the primary deposits.

The most notable localities for xenotime-bearing pegmatites and placers are in Brazil (Minas Geraes), Norway (Hittero near Arendal, Kragero in Telemarken), Sweden (Itterby), and elsewhere.

Apatite Group

The apatite group comprises compounds in which the cations are Ca^{2+}, Pb^{2+} and, in certain isomorphous admixtures, TR^{3+}, Y^{3+}, Mn^{2+}, and Sr^{2+}, the additional anions being F^{1-}, Cl^{1-}, $[OH]^{1-}$, O^{2-}, and $[CO_3]^{2-}$. group may be partly replaced by a weaker $[SiO_4]^{4-}$ anion, but in combination with a stronger $[SO_4]^{4-}$ anion (provided the cationic valences do not change).

Flour and Chlorapatites, $Ca_5[PO_4]_3F$ and Ca5[PO4]3Cl

"Apatao" is the Greek for / *deceive*. Formerly it was often mistaken for other minerals of prismatic or rod-like habit (beryl, diopside, tourmaline, etc.).

Chemical Composition: Fluorapatite: CaO 55.5 per cent, P_2O_5 42.3 per cent, F 3.8 per cent. Chlorapatite: CaO 53.8 per cent, P_2O_5 41 per cent, Cl 6.8 per cent. Fluorapatite is more widespread. Often F is present in addition to negligible amounts of Cl, OH, and CO_3 (carbonate-apatite). Sometimes isomorphous admixtures to CaO are Na_2O, rare earths, mainly

Ce_2O_3 (up to 5 per cent), MgO (up to a few per cent); insignificant amounts of Fe_2O_3, Al_2O_3, etc. System, hexagonal; symmetry, dipyramidal. Fluorapatite: a_0 = 9.36; c_0 = 6.88. Chlorapatite: a_0 = = 9.52; c_0 = 6.85.

Crystal Structure: The unit cell projected on the plane (0001) is rhombic, the fluorine ions occupying the corners of the rhombohedron. The Ca ions, according to N.V. Belov, are inside the trigonal prisms which form columns parallel to *c* axis. The columns are of two kinds: some of them have more complex structure consisting of three prisms in a given layer, with different orientation around the sixfold axis in each layer, the structural pattern, as a whole, thus acquiring hexagonal; other columns are simple, consisting of single trihedral prisms. All these columns are held together by the PO_4 tetrahedra which alternate with empty octahedra in the vertical direction. The fluorine anions are located in the centre of two layers of trigonal prisms. As for carbonate-apatite, according to Belov the $[PO_4]$ radical is partly replaced by $[Co_3F]$, which explains the increased fluorine content of this variety.

Habit: Frequently occurs in cavities as well-developed ingrowths or overgrowths of crystals in the form of hexagonal prisms and needles, less often as short-columnar or tabular forms. The most common faces are the prism (1010},pinacoid {0001}, dipyramids {1014}, {1121}, etc. The faces of the prisms may have vertical striations. In some hydrothermal iron ore deposits and pegmatites there also occur tubular ("hollow") hexagonal crystals of apatite. Aggregates. It is also widely distributed in granular, compact, cryptocrystalline, cross veined, and earthy masses. Widespread in sedimentary rocks in diverse concretionary forms, usually containing numerous tiny inclusions of other minerals (quartz, glauconite, calcite, etc.). These accumulations are commonly named *phosphorites* (phosphate rocks). Occurs as pseudomorphs after the bones of animals, pieces of wood, often with all the details of structure faithfully preserved, as recognised under the microscope.

Colour, Colourless (transparent), white, more often pale

green to emerald-green, light blue, yellow, brown, violet. *Lustre*, vitreous and greasy on fracture. Pure fluorapatite: n_y = 1.633 and n_z = 1.629.

Hardness, 5. Brittle. Fracture uneven, sometimes conchoidal. *Cleavage*, {0001} imperfect. *Specific gravity*, 3.18 to 3.21.

Diagnostic Features: Crystals are characterised by hexagonal prismatic habit. Are distinguished from similar crystals of beryl and aquamarine by lower hardness.

Fusible with difficulty, even in thin fragments in the blowpipe flame. Powder moistened with sulphuric acid colours the flame bluish green. Dissolves in nitric, hydrochloric and sulphuric acids. Nitric solution with ammonium molybdate gives the phosphorus reaction.

Genesis and Occurrence: Being one of the latest *magmatic* minerals, apatite occurs as microscopic grains in many igneous rocks in alkaline rocks (nepheline-syenites), occasionally as granular masses in association with silicates (nepheline, sphene, zircon, vesuvianite, etc.). Such, for instance, is the famous deposit in the *Hibiny mountains* on the Kola Peninsula.

In large rod-like or prismatic crystals it is found in many *pegmatites* of acid and alkaline igneous rocks. Excellent crystals of apatite are found in the *Slyudyanka* deposit near Lake Baikal. Sometimes occurs in *contact-metasomatic* formations in paragenesis with various minerals.

Is also found in some *hydrothermal* vein deposits as an accessory of such minerals as cassiterite, fluorite, etc. In Alpine-cleft veins short-columnar and even flat crystals of apatite are observed.

In the process of *weathering* of phosphorus-enriched limestones in karsts and leaching cavities there are formed accumulations of brown phosphorites with a concentrically-banded structure similar to that of agates.

Many large deposits of calcium phosphates are formed in marine sedimentary rocks through a complex biochemical process and are confined to definite stratigraphic horizons.

The phosphorites in the form of concretions and nodules 43) of most diverse shapes, less often massive, occur in argillaceous and glauconitic sands and sandstones. Phosphorites may also form out of guano, the excrements of sea fowl on desert coasts composed of limestones or other calcium-bearing rocks.

Large phosphorite deposits are found in the Ukraine. A notable deposit is located in the South-Russian depression where a phosphorite-bearing horizon outcrops intermittently in the Upper Cretaceous sediments over a vast area from Podolia (Western Ukraine) to the Volga shores (near Saratov); its northern boundary intrudes into the Kursk Region, and the southern boundary runs across the southern rim of the Donets Basin.

The phosphorite is heavily mixed with sand and occurs in nodules or relatively thin beds. Another major deposit of massive laminated high-grade phosphorite has been discovered recently in the *Kara-Tan* mountains (Southern Kazakhstan) in Lower Paleozoic strata. Massive phosphorite of varying dark hues sometimes has an oolitic structure.

The most important foreign localities are the huge beds of phosphorites (phosphate rock) of Permian age which occur in the *United States* (Idaho, Wyoming, Utah, Montana), and the deposits in Lower Tertiary strata along the coast of *Algeria, Tunisia* and *Morocco.*

Uses: The principal use of apatite and phosphorites (up to 90 per cent) is for the production of fertilizers (various superphosphates and thermophosphates). Apatites are a source of phosphoric acid and various salts as well as of phosphorus, an important material for safety matches. Fluorine-rich apatites yield hydrofluosilicic acid, as a by-product. Apatite also goes to make the so-called bone china.

Pyromorphite, $Pb_5[PO_4]_3Cl$

"Pyros" is the Greek for fire; "morphe", form.

Chemical Composition: PbO 82.0 per cent, P_2O_5 15.4 per cent,

C 12.6 per cent. Sometimes contains CaO (up to 8-9 per cent), and As_2O_5 (up to 4 per cent). Occasionally an admixture of chromium is observed (probably as cr_2O_3) which colours the mineral bright red; rarely V_2O_5.

System, hexagonal; *symmetry,* dipyramidal, crystal structure is analogous to that of apatite. Habit. Usually occurs in prismatic or columnar crystals.

May occur in barrel-shaped, acicular, often in parallel-intergrown crystals. Aggregates, apart from common druses of small crystals, often occurs as crystalline reniform and globular formations.

Colour, Various tinges of green, yellow, and brown. Less common are varieties coloured bright red or orange-yellow by chromium. The colouring is often banded. Streak, white, sometimes yellowish. Lustre, adamantine, greasy. n_y—2.050 and n_z = 2.042.

Hardness, 3.5 to 4.0. Brittle. Practically no cleavage. *Specific gravity,* 6.7 to 7.1, diminishing to 5.9 for calcium-rich varieties.

Diagnostic Features: Pyromorphite crystals differ from apatite by specific gravity and adamantine lustre. Often indistinguishable from mimetesite and vanadinite without chemical tests. Ignited with soda, gives a lead globule. On charcoal gives a yellow coating of PbO_2. Soluble in nitric acid, and also in KOH (if it contains no lime).

Genesis and Occurrence: Pyromorphite is a mineral of almost exclusively *exogenous* origin, being formed in the oxidised zones of sulphide lead and lead-zinc deposits (as druses of small crystals and crystalline crusts in leaching cavities). The source of phosphoric acid are probably the surface waters reacting with lead minerals.

Often pseudomorphous after cerussite and galena. Reverse cases, i.e., replacement of pyromorphite by galena on the periphery of crystals, are also observed under certain specific conditions. Pseudomorphous after pyromorphite are also vanadinite, vauque-linite, wulfenite, calamine, chalcedony, and calcite. Has been recognised from certain low-temperature vein deposits as an *endogenous* mineral. Lastly, pyromorphite often occurs as an accessory mineral in the oxidised zones of many deposits. In the USSR, it is found in the oxidised zones of the *Berezovskoye* gold deposit in the Urals (together with crocoite).

Excellent crystals of pyromorphite with an admixture of a negligible quantity of As_2O_5 have been recorded from *Shilka* and *Zerentuy* deposits (Eastern Trans-Baikal area) at *Kyzyl-Espe* (Kazakhstan), and elsewhere.

Uses: Together with the other lead minerals of the oxidised zone pyromorphite is a source of lead. It rarely occurs in considerable amounts.

Mimetesite. Pb $[AsO_4]_3Cl$

"Mimetes" is the Greek for imitator -reference to its similarity to pyromorphite.

Chemical Composition: PbO up to 74.9 per cent, As_2O_5 up to 23.2 per cent, Cl up to 2.4 per cent. Often contains P_2O_5, Sb_2O_5, also CaO up to 10-14 per cent, and occasionally BaO and Cr_2O_3, as isomorphous admixtures.

System, hexagonal; *symmetry,* dipyramidal. Habit, as a rule, in prismatic, often acicular forms. Barrel-shaped, tabular and dipyramidal crystals are also known. Aggregates. In cavities, as druses and rosettes of small crystals, crusts, more seldom as sinter-like and earthy masses.

Colour, Pure varieties are colourless and transparent. More often honey-yellow, brown or green. *Streak,* white. *Lustre,* adamantine, greasy. n_y = = 2.135 and n_z = 2.120.

Hardness, 3.5. Brittle. Fracture conchoidal, uneven. Practically no *cleavage. Specific gravity,* 7.19 to 7.25.

Diagnostic Features: May be positively distinguished from pyromorphite and vanadinite only by chemical tests. Mimetesite crystal. Before the blowpipe on charcoal fuses with more difficulty than pyromorphite, and giving off vapours of oxidised arsenic with a garlic smell gives a lead globule. Dissolves in sulphuric and hydrochloric acids with separation of $PbSO_4$ and $PbCl_2$ in the form of a poorly soluble precipitate.

Genesis and Occurrence: Like pyromorphite, occurs in the *oxidised zones* of sulphide lead-zinc deposits, but much less often than the former.

The sources of arsenic in this case are probably oxidised arsenides, sulphoarsenides, and sulphoarsenites, such as arsenopyrite and grey ores present as impurities in primary lead ores. Reference has also been made in literature to the formation of endogenous mimetesite in a low-temperature *hydrothermal* deposit at Lilli in the Sica-Sica province (Bolivia) in veinlets with quartz, chalcedony, calcite, etc.

In the USSR, reniform masses and crystals have been found in a number of lead-zinc deposits in Eastern Trans-Baikal area (Nerchinsk district): Trekhsvyatitelsky, Spasskoye, Klichkinskoye, Kadainskoye deposits, and also Kyzyl-Espe (Kazakhstan), and elsewhere.

Outside the Soviet Union, the deposit of Tsumeb (South-Western Africa) is notable for its large well-formed crystals; other localities are Chihuahua (Mexico) and Tintic in Utah (U.S.A.).

Vanadinite, $Pb_5[VO_4]_3Cl$: Named after the important element vanadium contained in the mineral.

Chemical Composition: PbO 78.3 per cent, V_2O_5 19.3 per cent, Cl 2.4 per cent. Sometimes contains negligible amounts of P_2O_5, As_2O_5, etc.

System, hexagonal; *symmetry,* dipyramidal. Habit, crystals are, as a rule, small and of prismatic or acicular habit. Often in small crystals in the form of hollow prisms or narrow dipyramids forming crystal druses in cavities, and parallel intergrowths of prismatic crystals. Occurs in reniform aggregates with a radial-fibrous or fibrous structure.

Colour, yellow, brown, sometimes red. *Streak*, white or pale yellow. *Lustre*, adamantine, greasy. $n_y = 2.354$ and $n_z = 2.299$.

Hardness, 3.0. Brittle. Fracture uneven. *Cleavage*, none. *Specific gravity*, 6.66 to 7.10.

Diagnostic Features: Vanadinite is less hard than pyromorphite and mimetesite. Blowpipe and chemical tests are required, however, for more positive identification.

In the blowpipe flame on charcoal decrepitates, readily fuses into a black shiny mass, which in the reducing flame gives a lead globule, and a yellow coating of $PbCl_2$ on charcoal. In the reducing flame the salt of phosphorus bead is reddish yellow, turning purely green (vanadium reaction) on cooling.

The powdered mineral, when dissolved in a drop of dilute nitric acid on a slide and completely evaporated, gives a dark-red residue (other minerals of the group give a white or grey residue).

Genesis and Occurrence: Like pyromorphite and mimetesite, is formed in the oxidised zones of lead-zinc sulphide deposits, mostly in carbonate rocks. Occasionally pseudomorphous after pyromorphite. In the *Suleyman-Sai* deposit (Southern Kazakhstan) large quantities of vanadinite have been found as well-developed crystals, earthy and friable masses and crusts coating remnants of unoxidised finegrained galena.

Notable foreign localities are at *Broken Hill* (Northern Rhodesia) and *Lamentos* (Mexico), where vanadinite, descloizite, and wulfenite characteristically accumulate in the lower parts of the oxidised zone.

Uses: When found in quantity, vanadite-bearing ores are an economic source of vanadium.

Summary: It follows from the above description that the minerals of this group, along with common properties, have considerable differences determined by their chemical composition. A comparison of their principal characteristics-is given below.

Basic Characteristics of the Apatite Group Minerals

Mineral	*Specific gravity*	*Hardness*	*Refractive index* $n_{\parallel}$	$n_{\perp}$	*Lustre*
Fluorapatite, . . $Ca_3[PO_4]_3F$	3.2	5	1.633	1.629	Vitreous Adaman- tine
Pyromorphite, $Pb_5[PO]_3Cl$	6.9	3.5-4	2.050	2.042	
Mimetesite, $pb_5[As\ O_4]_3Cl$	7.2	3.5	2.125 ;	2.120	Ditto
Vanadinite, $Pb_6[VO]_3Cl$.	7.0	3.0	2.354	2.299	Ditto

A comparison of the properties of pyromorphite, mimetesite and vanadinite shows that with increasing replacement of P in the $XO^{\pm}$, complex anion by As and V, the hardness of the mineral decreases and the refractive index increases. The differences in specific gravity are not conspicuous since the content of heavy metal (lead) is strongly predominant over the other elements.

The considerable distinctions between the properties of apatite and the minerals just reviewed are quite natural.

Amblygonite-triplite Group

Amblygonite, Lial [PO$_4$]F: System, triclinic. Occurs in large, poorly formed crystals. Colour, white with greenish, bluish, yellowish, and other tinges. Translucent. Lustre, vitreous; pearly on cleavage surfaces. *Hardness,* 6. *Brittle.* Cleavage, {001} excellent. *Specific gravity,* 2.98 to 3.15.

Before the blowpipe fuses readily with intumescence. Colours the flame red (presence of lithium). With powdered magnesium gives the phosphorus reaction. Soluble in sulphuric acid. Has been recorded from pegmatite dikes associated with other lithium-bearing minerals at Estremadura (Spain), Herbon, Maine (U.S.A.), and elsewhere.

Lithiophilite, Li (Mn, Fe) PO$_4$ Rare

System, orthorhombic. *Colour,* pale pink, yellow, red-brown, light blue. *Streak,* white or faintly coloured. *Lustre,* vitreous.

Hardness, 4.5 to 5. *Cleavage,* {001} perfect, {010} less perfect. *Specific gravity,* 3.5.

Fuses readily before the blowpipe flame colouring the flame red. Soluble in acids. Gives reactions for Mn, Fe, and P. Has been recorded from pegmatites with quartz, beryl, and lithium minerals at Branchville, Connecticut (U.S.A.), and elsewhere. In the oxidised zone is subject to alteration.

Tbiplite, (Mn, Fe)$_2$ [PO]F:

Contains isomorphous admixtures of Ca and Mg. *System,* monoclinic. Sometimes massive. Colour, brown or pink. *Hardness,* 4 to 5. *Cleavage,* poor. *Specific gravity,* 3.44 to 3.87.

Before the blowpipe fuses readily into a black magnetic globule; with borax in the oxidising flame gives a violet bead (Mn). With powdered Mg gives the phosphorus reaction. Soluble in hydrochloric acid.

Occurs in pegmatite dikes in association with quartz, beryl, apatite, fluorite, and other minerals, and also in some hydrothermal veins. In the USSR, has been found in hydrothermal deposits in Buryatia in association with quartz, rhodochrosite, etc. Is subject to alteration in the oxidised zone.

Hydrous Phosphates, Arsenates, and Vanadate

VIVIANITE GROUP

The minerals of this group are hydrous normal phosphates and arsenates with 8 water molecules hydrating the cations of iron, nickel, and cobalt.

Vivianite, Fe_2 $[PO_4]_2.8H_2O$

Chemical composition. FeO 43.0 per cent, P_2O_5 28.3 per cent, H_2O 28.7 per cent. Often contains ferric oxide which has developed from the ferrous form on further oxidation.

System, monoclinic; *symmetry,* prismatic. *Habit,* commonly prismatic 263), rod-like, acicular. Aggregates. Often occurs in radial-fibrous, stellate, reniform, globular aggregates, and also as earthy masses (blue iron earth).

Colour, Unaltered vivianite is light-coloured or even colourless and transparent; on partial oxi *.010* dation in the air becomes greyish blue or greyish green, dark blue to black-blue. *Streak,* unoxidised varieties colourless. Oxidised varieties bluish white, blue or brown. *Lustre,* vitreous; pearly on cleavage surfaces. n_x = 1.633, n_y = 1.603, and n_z = 1.579.

Hardness, 1.5 to 2. Brittle. *Cleavage,* {010} ex-*263*. *Specific gravity,* 2.95.

Diagnostic Features: Partly oxidised varieties may be identified by greenish-blue or blue colour, streak, and low hardness. Also very characteristic are radial-acicular, rod-like, stellate aggregates, quite frequently occurring in cavities in accumulations of iron hydroxides, in fossil bones and shells.

In the reducing flame turns red and fuses into a grey lustrous magnetic globule. With salt of phosphorus gives the characteristic iron bead. Heated in the glass tube, yields much water, the reaction of which is neutral, intumesces, and turns grey with red spots. Readily soluble in hydrochloric and nitric acids.

Genesis and Occurrence: Vivianite is formed through exogenous processes in a reducing environment. Its phosphorus is often derived from organic remains. Commonly found in phosphorus-rich sedimentary iron ores, and also in peat bogs (as an earthy variety) associated with siderite and other ferrous oxide minerals. Is subject to alteration in the oxidised zone.

Vivianite occurs in many deposits in the USSR Well-known are crystalline radiated aggregates in shells and cavities in the brown iron ores of the Kerch Peninsula (the Crimea) and on the Taman Peninsula. Earthy vivianite is widely distributed in pig iron ores, e.g., in the Moscow Region.

Uses: When found in quantity, is a source of cheap blue pigment.

Annabergite, $Ni_3[AsO_4]_2$-$8H_2O$

Named after the place of occurrence (Annaberg in Saxony). Synonym: nickel bloom.

Chemical Composition: NiO 37.5 per cent, As_2O_5 38.5 per cent, H_2O 24.0 per cent. Sometimes contains a negligible isomorphous admixture of Co, Fe, Mg, rarely Ca.

System, monoclinic; *symmetry,* prismatic. Crystals capillary and rare. Usually observed as earthy masses.

Colour, apple-green to dark green. *Streak,* slightly greenish. *Lustre,* vitreous. $n_x = 1.687$, $n_y = 1.658$, andra$_z$ = 1.622. Hardness, 2.5 to 3. Brittle. Cleavage, {010} perfect. *Specific gravity,* 3.0.

Diagnostic Features: Characteristic green colour and association with nickel arsenides such as niccolite and chloanthite, from which it usually derives.

Fuses in the reducing blowpipe flame evolving arsenic and giving a greyish-black globule. Readily soluble in acids. With dimethyl glyoxyme gives a vivid nickel reaction.

Genesis and Occurrence: Confined exclusively to the oxidised zones of ore deposits, the primary ores of which contain nickel arsenides. As a rule, they are replaced directly by annabergite.

Occurs in quantity in nickel-arsenide deposits at Annaberg and Schneeberg in Saxony. Finds in the Urals, North Caucasus, and elsewhere in the Soviet Union are of mineralogical interest only.

Erythrite, $Co_3[AsO_4]_2.8H_2O$

"Erythros" is the Greek for red. Synonym: cobalt bloom.

Chemical Composition: CoO 37.5 per cent, As_2O_5 38.5 per cent H_2O 24.0 per cent. Occasionally contains isomorphous admixtures of Ni, Mg, Fe, Ca, and also foreign matter as mechanical impurities.

System, monoclinic; *symmetry,* prismatic. Rare crystals are small. *Habit,* acicular or thin-platy. Usually in earthy masses.

Colour, crimson to peach-red, dark pink. Streak, pale red. *Lustre* of crystals, vitreous; pearly on *cleavage* surfaces. $n_x = 1.701$, $n_y = 1.663$, and $n_z = 1.629$.

Hardness, 1.5 to 2.5, for earthy masses 1. *Cleavage,* {010} excellent. *Specific gravity,* 2.95.

Diagnostic Features: Readily identifiable by pink or red colour, earthiness, and association with cobalt arsenides: cobaltite and smaltine.

Before the blowpipe fuses into a grey bead giving an arsenical (garlic) odour, and colouring the flame light blue. Very characteristic are the deep blue beads of borax and salt of phosphorus. In the closed tube yields much water. Soluble in hydrochloric acid colouring the solution pink-red.

Genesis and Occurrence: Like annabergite, occurs in the oxidised zones of the deposits whose ores contain cobalt arsenides. May be pseudomorphous after smaltine. Is found in the same deposits as annabergite, since arsenides of nickel and cobalt mostly occur together. However, certain deposits contain cobalt arsenides only and, therefore, erythrite is not accompanied by annabergite. Such are the deposits of Dashkesan (Azerbaijan), Ak-Jilga (the Alai range in Central Asia), etc.

Scobodite Group

Scorodite, Fe [AsO_4] $2H_2O$

"Scorodon" is the Greek for garlic. Has been named for its characteristic garlic odour given off when it is struck with a hammer.

Chemical Composition: Fe_2O_3 34.6 per cent, As_2O_5 49.8 per cent, H_2O 15.6 per cent. Varieties containing up to 7.1 per cent of Al_2O_3 known as *alumoscorodite*. Massive scorodite is often contaminated with foreign matter.

System, orthorhombic; *symmetry,* dipyramidal. *Habit,* dipyramidal, sometimes in combination with prisms. Frequently observed in finely-crystalline druses. Usually in compact earthy masses. Colour, greenish white, apple-green, brownish. Scorodite crystal. grey, sometimes white (in cryptocrystalline earthy masses). *Pitticite,* an impure partly hydrolysed variety coloured by iron hydroxides, is reddish brown. n_x = 1.79 to 1.81, n_y = 1.77 to 1.79, and n_z = 1.76 to 1.78.

Hardness, 3.5. Brittle. Fracture, uneven. *Cleavage,* {100}, (001), and {201} imperfect. *Specific gravity,* 3.3.

Diagnostic Features: Characteristic pale-green hues of minute pyramidal crystals in cavities or of continuous, often dull, masses. Highly aracteristic is the association with residual unreplaced arsenorite.

Readily fusible before the blowpipe, colouring the flame light blue id giving off the garlic odour. On charcoal with soda gives a black agnetic slag. Readily soluble in hydrochloric acid and, to some extent, in nitric acid. A KOH solution rapidly colours scorodite powder dish brown.

Genesis and Occurrence: Scorodite forms sometimes considerable asses in the oxidised zones of arsenopyrite-rich deposits. Usually velops metasomatically directly after arsenopyrite and lollingite.

In the USSR, occurs in many arsenopyrite-rich sulphide ore posits. In the *Brick-Mulla* deposit (northeast of Tashkent) scorobe has been found in large quantities both as granular massive and cryptocrystalline gel-like formations with a waxy lustre of various colours: white, all shades of green, brown, and even black. Scorobe also occurs at *Takeli* in the Karamazar ridge (Tajikistan), at *jchkar* and *Jetygar* (Southern Urals), *Zapokrovskoye* (Eastern Trans-ural area), etc.

Uranite Group

This group includes the rather numerous hydrous basic phosphates, senates, and vanadates which are mostly binary salts of divalent etals, such as Cu^2+ and Ca^2+, followed by Mg^2+, Fe^{2+}, Mn^{2+}, Ba^{2+} and K^{1+} with U^{6+}; the latter is believed to form a $[UO_2]^{2+}$ tionic group with additional O^{2-} anions. The most common among em are compounds with eight water molecules. It should be men-made that by far not all mineral species of the group occurring in ture have been identified. The important characteristic of the uranites is perfect unidirectional cleavage, in which respect they resemble micas.

Their crystal structures are characteristically laminated. The minae are probably double sheets composed of XO_4 tetrahedral oups, linked by U^{6+}cations having sixfold oxygen coordination. Cu, Ba cations and H_2O molecules may be located between such layers.

The water-rich varieties on exposure to air are dehydrated owing the low vapour pressure of the environment, the process being specially intensified by heating. This involves changes in the physical properties, especially of specific gravity and refractive indices.

The minerals of the group are distinguished by bright-yellow or green colour, iridescent chatoyancy on cleavage surfaces, comparative low hardness, good solubility in acids, and strong radioactivity, these compounds are easily produced artificially from cold solutions appropriate composition and concentration.

Lastly, practically all these minerals occur under the same conditions, being the products of the oxidation of uranium deposits. These minerals are of economic importance as a source of uranium, radium, and, sometimes, vanadium.

Torbernite, $Cu[UO_2]_2[PO_4]2\text{-}12H_2O$

Chemical Composition: According to the analytical data available: CuO 7.73 per cent, UO_3 57.50 per cent, P_2O_5 14.50 per cent, H_2O 20.30 per cent. Very closely agrees with the theoretical composition. At 45°C four water molecules disappear, the mineral changing to metatorbernite.

System, tetragonal; *symmetry,* ditetragonal dipyramidal. *Habit,* crystals are small, well-formed, and tabular.

Intergrowths with autunite have been observed. Torbernite colour, emerald green. *Lustre,* vitreous, crystal, pearly on cleavage surfaces. n_y = 1.590 to 1.592, and n_z = 1.576 to 1.582.

Hardness, 2 to 2.5. *Cleavage*, {001} perfect. *Specific gravity*, 2.3 to 3.6. Strongly radioactive.

Diagnostic Features: Characteristic are perfect cleavage (micaceous), and bright-green colour. Torbernite may be distinguished from other green minerals by optical constants and chemical tests. Before the blowpipe fuses into a black bead. With soda in the reducing flame yields a copper globule. Dissolves in nitric acid; when ammonia is added, the solution turns blue (copper) giving a yellow precipitate.

Genesis and Occurrence: Forms in the oxidised zones of pegmatites and hydrothermal and also sedimentary and other uraniferous deposits. Usually is found in small quantities on the walls of cracks and leaching cavities, often on limonite, sometimes in association with autunite and other secondary uranium minerals. Torbernite is a relatively widespread mineral.

Autunite, $Ca[UO]_2[PO_4]_2$ $8H_2O$

Named after a locality in France.

Chemical Composition: Cap 6.1 per cent, UO_3 62.7 per cent, P_2O_5 15.5 per cent, H_2O 15.7 per cent. Sometimes contains impurities of BaO, MgO, Fe_2O_3, etc. Autunite may also contain actinium and polonium.

System, tetragonal; *symmetry*, ditetragonal dipyramidal.

Habit, crystals are thin-tabular, micaceous, square. In crystal terfacial angles are close to those of torbernite.

Druses of small crystals or scaly aggregates are sometimes observed.

Colour, green, greenish yellow, sulphur-yellow. *Lustre*, on *cleavage* surfaces pearly. $n_x = 1.594$, $n_y = 1.590$, and $n_z = 1.571$.

Hardness, 2 to 2.5. Brittle. *Cleavage*, {001} highly perfect, {100} and {010} less perfect. Specific gravity, 3.05 to 3.19. Strongly radioactive. Luminescent. In ultraviolet radiation fluoresces yellow-green.

Diagnostic Features: May be positively distinguished from other similarly coloured uranites by spectral and chemical analysis.

Slightly intumesces before the blowpipe flame and fuses into a black globule. Colours the flame orange-red (Ca). In the closed tube, yields much water. With borax shows the reaction for uranium. Soluble in nitric acid colouring the solution green.

Genesis and Occurrence: Like all the uranites, is formed in the oxidised zones of uranium deposits.

Autunite occurs in a fairly numerous deposits: Autun in the Loire department (France), Schneeberg, Johanngeorgenstadt in Saxony (Germany), Redruth in Cornwall (Britain), pegmatites of Madagascar, and elsewhere.

Tyuyamunite, $Ca[UO_2]_2[VO_4]_2$ $8H_2O$: Discovered by Soviet scientist K. A. Nenadkevich at Tyuya-Muyun (Ferghana).

Chemical Composition: CaO 5.87 per cent, UO_3 59.96 per cent, V_2O_5 19.06 per cent, H_2O 15.11 per cent. Sometimes contains as negligible impurities Na_2O, K_2O, MgO, CuO (up to 4 per cent), SiO_2, etc.

System, orthorhombic, *symmetry,* rhombic dipyramidal. Habit, thin-tabular. Common forms: {100}, {010}, {110}, {120}, {001}, {101}, and {111}. Aggregates. Thin-scaly earthy masses and also as coatings and crusts in cavities.

Colour, bright yellow (canary), sometimes with greenish or slightly orange tinge. *Lustre,* strong, pearly on *cleavage* surfaces. $n_x = 1.895$, $n_y = 1.870$, and $n_z = 1.670$.

Hardness, 1. Brittle. *Cleavage,* {001} perfect, {010} and {100} distinct. *Specific gravity,* 3.68 (varies depending on water content). Strongly radioactive.

Diagnostic Features: Distinguishable from other uranites only by chemical tests and optical constants. In finely-dispersed masses resembles the bright-yellow efflorescence of certain iron sulphates. Is distinguished from the latter by poor solubility in water and radioactivity.

Readily fusible in the blowpipe flame. With borax gives a distinct reaction for uranium (in the absence of copper). In the closed tube, yields much water. Readily soluble in acids. With hydrogen peroxide, gives a reaction for vanadium. In appropriate solutions and concern-resulting in the formation of carnotite.

Genesis and Occurrence: Together with other exogenous minerals, tyuyamunite occurs in the oxidised zone of uranium-bearing deposits. It is also formed in the presence of organic matter with which vanadium minerals are often associated in general. Tyuyamunite may quite possibly be formed after carnotite in sedimentary rocks through the action of waters containing calcium bicarbonate. The possibility of potassium being replaced by calcium has been confirmed experimentally.

In the USSR, usually occurs as earthy masses associated with gypsum in cracks in sedimentary rocks containing organic remains. A notable foreign locality are the vanadium-bearing sandstones of Colorado (U.S.A.). Also found in the calcareous sandstones of *Katanga* (Congo), *Radium Hill* (Australia), and many other places.

TURQUOISE GROUP

The group includes pentahydrous basic phosphates of Cu, Al, and Fe^{3+} crystallising in the triclinic system. We shall limit the present discussion to turquoise.

Turquoise, $CuAl_6[PO_4]_4[OH]_8.5H_2O$

Synonym:kallait (the old name for turquoise). A variety rich in iron (Fe_2O_3 20 to 21 per cent) is called *rachleite*.

Chemical Composition: CuO 0.57 per cent, Al_2O_3 36.84 per cent, P_2O_5 34.12 per cent, H_2O 19.47 per cent. May contain various impurities.

System, triclinic; *symmetry*, pinacoidal. Commonly occurs in crypto-crystalline reniform masses or crusts, veinlets, and irregular masses.

Colour, sky-blue, apple-green, greenish grey. *Lustre*, waxy. n_x = - 1.65, n_y = 1.62, and n_z = 1.61.

Hardness, 5 to 6. Rather brittle. *Cleavage*, {001} perfect, {010} distinct. Fracture, slightly conchoidal. *Specific gravity*, 2.60 to 2.83.

Diagnostic Features: Characteristic colour and waxy lustre. Sometimes, however, it is necessary to resort to chemical tests to distinguish it from similar chrysocolla and other copper minerals.

Before the blowpipe decrepitates turning brown. Colours the flame pale green. In the closed tube, yields much water. With borax and salt of phosphorus gives the reaction for copper. Soluble in acids. Gives the phosphorus reaction.

Genesis and Occurrence: Turquoise is formed in the zone of weathering, often together with limonite, through the action of cupriferous surface solutions on rocks containing alumina (feldspars) and phosphorus (apatite). A variety known as bone turquoise or odontolite is represented by fossil bones and teeth coloured blue. The organic genesis of such turquoise is readily visible in polished sections under the microscope.

The best turquoise has come for centuries from the *Madan* deposit near Nishapur (Iran) where it was formed together with limonite as irregular masses and thin veinlets in weathered igneous rocks, trachyte. It is from there that gem turquoise has come to Europe via Turkey. Other localities worth of mention are: *Vadi-Magara* (Sinai Peninsula), *Kara-Tyube* (south of Samarkand), etc.

Uses: Best coloured varieties (sky-blue) are used as gemstonesu Varieties of inferior quality may be coloured artificially.

Borates

General. This class includes salts of boric acids such as orthoacid HBO_3, metaboric acid, HBO_2, and the so-called polyboric acids.

These hypothetical polyboric acids are deduced by subtracting a certain number of H_2O particles from the corresponding number of normal orthoacid particles. The most common in nature are salts of the following boric acids:

$H_3BO_3.H_2O \rightarrow HBO_2$ $5H_3BO_3.7H_2O \rightarrow HB_5O_8$

$4H_3BO_3.$ $5H_2O \rightarrow H_2B_4O_7$ $6H_3BO_3.$ $7H_2O \rightarrow H_4B_6O_U$

$5H_3BO_3.$ $6H_2O \rightarrow H_3B_5O_9$ $7H_3BO_3.$ $8H_2O \rightarrow$ $H_5B_7O_{13}$ etc.

Borates may contain Al^{3+}, Fe^8+, and Mn^3+ cations which are known to form salts in combination with divalent cations of small ionic radii (Mg^{2+} and Fe^{2+}). The class also includes acidic and basic orthosalts of Mg^{2+}, which in one case are combined with Ti^{4+}. Calcium occurs in binary salts only.

All these *orthoborates* occur exclusively as anhydrous compounds. Normal orthoborates are insoluble not only in water, but even in acids (or are decomposed by them with difficulty), fusible at high temperatures and have very high or high hardness.

Polyborates besides the Mg^{2+} cation, may contain larger Ca^{2+} and Na^{1+} cations, and occur almost invariably as hydrous salts. Polyborates of Na^{1+} and other strong cations are readily soluble in cold water, while binary borates of Na^{1+} and Ca^{2+} are soluble in hot water. Anhydrous polyborates of Mg (e.g., boracite) are gradually hydrated in water at room temperature, in which respect they differ from orthoborates.

The behaviour of boron in the course of natural processes shows some other interesting features as well. Whereas at low temperatures B_2O_3 is replaced by CO_2, at high temperatures, the process is reversed. Borates may be replaced by carbonates in the course of weathering which has been observed at the Inner deposit. In contact-metasomatic deposits, on the other hand, there have been found borates formed by replacement in limestones.

Boron is a rather mobile element, being transported by water solutions containing such constituents as Cl, OH, and especially F which has a strong chemical affinity for boron. Therefore, the concentration and formation of boric compounds take place in the *residual products* of geological processes. Borates are closer to silicatesin a number of crystallochemical properties than to any other oxysalts. Borates that contain simple anionic groups $[BO_3]^{3-}$ and $[BO_4]^{5-}$ in their crystal structures differ but slightly from typical oxysalts, including the orthosilicates with their $[SiO_4]^{4-}$ groups. In contrast to salts of other types, however, borates, as well as silicates may contain more complex anionic $[BO_3]^{3}$-groups interlocked to form common triangular apexes 268), which gives rise to the so-called extended anions in the form of negatively charged chains, layers, frameworks, etc., characteristic of polyborates.

Anhydrous Berates

Anhydrous borates are represented mostly by salts of orthoboric acid (normal, acid, and basic). There are few minerals that may be classified as meta- and polyborates.

Ascharite, $MgHBO_2$ or $2MgO\ B_2O_3\ H_2O$

Named after a geographical location: Ascharix, a former Roman province (Aschersleben in Saxony). Synonym: szaibelyite (the term used in the USA).

Chemical Composition: MgO 47.91 per cent, B_2O_3 41.38 per cent, H_2O 10.71 per cent. May contain Mn and negligible F.

System: Orthorhombic occurs in friable chalk-like, more seldom in compact masses of parallel-fibrous or thin-acicular structure.

Colour, white. Lustre, vitreous, in friable masses dull. *Hardness,* 3 to 3.5. In friable chalk-like masses smears the fingers. *Specific gravity,* 2.65.

Diagnostic Features: Ascharite may be identified tentatively when it is found in paragenesis with other borates. Positive recognition is possible only by measurement of optical constants and by chemical analyses.

Before the blowpipe fuses into enamel and strongly colours the flame green (presence of boron). Insoluble in water. Soluble with difficulty in acids. Yields the bulk of its water at temperatures above 800°.

Genesis and Occurrence: Ascharite occurs most often as a secondary mineral in salt-bearing sedimentary deposits of borates. It is formed in the conditions of a hot arid climate, as a product of the natural reworking and dehydration of other, chiefly hydrous, boracic compounds. May accumulate metasomatically in clays. In leaching cavities occurs as colloform reniform masses of radial or acicular structure.

Cases are known, when it has been found in contact-metasomatic deposits in limestones or dolomites associated with serpentine (magnesium hydrosilicate), magnetite, and other minerals.

Notable foreign localities are Aschersleben and Neustassfurt (Saxony), Rezbanya (Rumania), Nevada (U.S.A.).

Uses: When found in large quantities together with other borates, ascharite is of economic importance.

Borates are a source of boric acid, which has various uses. The chemical industry uses boric acid to make different boric salts. Together with borax it is used in enamels for iron ware (its expansion coefficient approaches that of the metal), for impregnating candle wicks to keep them upright when burning, in tanning, pharmaceutics, etc.

Furthermore, borax is used in making lamp-glass and other kinds of glass subject to temperature fluctuations such as "pyrex" (B_2O_3 11.8 per cent), for laboratory utensils. Since borax dissolves metal oxides, it is indispensable as a flux which prevents formation of oxides on the surface of soldered joints.

Borax is also added to paper pulp in the manufacture of high-grade glossy and non-inflammable papers, and articles therefrom. Boron compounds are also important fertilizers; correctly dosed, they help to balance plant nutrition with potassium salts, and thus to increase crop yields.

Of late, boron has been used extensively in highly efficient jet fuels. Boron carbides (B_6C, B_4C, and B_3C) are good abrasives. Borides of W, Ti, and Ta are very hard and refractory. Boron steel is used in atomic piles. Boracite, $Mg_3B_7O_{13}Cl$.

Chemical Composition: B_2O_3 62.1 per cent, MgO 30.7 per cent, Cl 8.1 per cent.

System. Pseudocubic, above 265° becomes cubic. Below 265°C converts to an orthorhombic modification. Space group of the high-temperature modification. Habit, cubic or octahedral. Also occurs as continuous close-grained marble-like masses.

Colour, white with greyish, yellowish or greenish tinge. *Lustre,* vitreous, strong. $n_x = 1.673$, $n_y = 1.667$, and $n_z = 1.662$.

Hardness, 7. *Cleavage,* none. Fracture conchoidal. *Specific gravity,* 2.95.

Diagnostic Features: Characteristic crystals in the cubic system, which in thin polished sections under the microscope show optical anisotropy and complex texture. High hardness (7) and rather high specific gravity are also distinctive features of boracite.

Before the blowpipe fuses rather easily into a white enamel globule, colouring the flame green. Insoluble in water, dissolves slowly in hydrochloric acid. On weathering absorbs water and becomes fibrous (evidently altering to a different mineral).

Genesis and Occurrence: Occurs in salt deposits with carnallite, sylvite, halite, gypsum, anhydrite, and other minerals. Evidently forms in the process of metamorphism through dehydration of hydrous borates of magnesium. Is found at Stassfurt in Anhalt, at Luneburg in Hannover (Germany) and other deposits. Has commercial importance.

Carnotite, Kluojitvoji-shjo. Named after A. Garnot.

Chemical Composition: K_2O 10.44 per cent, UO_3 63.41 per cent, V_2O_5 20.16 per cent, H_2O 5.99 per cent. Sometimes contain as impurities Na_2O, MgO, CaO, a very negligible quantity of CuO, PbO, etc.

System, monoclinic; *symmetry,* rhombic prismatic. Usually in pulverulent masses and coatings.

Colour, bright yellow to greenish yellow. Lustre, strong; pearly on *cleavage* surfaces. $n_x = 1.950$, $n_y = 1.925$, and $n_z = 1.750$. *Hardness,* 2 to 2.5. Brittle. *Cleavage,* {001} perfect. *Specific gravity,* 4.46. Strongly radioactive.

Diagnostic Features: Has much in common with tyuyamunite, but differs from it in optical constants. Before the blowpipe fuses readily into a black globule and colours the flame pale violet (should be viewed through blue glass which filters out the yellow tint of sodium). With borax gives a uranium bead. Readily soluble in acid, with hydrogen peroxide gives the vanadium reaction. Its powder, when slightly moistened with hydrochloric acid, turns blood-red, this indicating the presence of vanadium.

Genesis and Occurrence: Occurs in the weathered zone of sedimentary rocks, chiefly in sandstones enriched in organic matter. First discovered in Jurassic vanadium-bearing sandstones in *Utah* and *Colorado* (U.S.A.).

Also found in the calcareous sandstones of *Katanga* (Congo), *Radium Hill* (Australia), and many other places.

Hydrous Borates

Almost all the hydrous salts reviewed here are polyborates of Mg, Na, and Ca.

Borax, $Na_2B_4O_7$-$10H_2O$

System, monoclinic. Sometimes occurs in large prismatic crystals. Also widespread as earthy masses.

Colour, Colourless but usually white with greyish, yellowish, bluish, or greenish tinges. *Lustre,* vitreous, greasy.

Diagnostic Features: Characteristic low hardness, frequent asbestos-like structure with silky lustre and insolubility in water. For positive identification, chemical tests and measurements of optical constants are necessary.

Swells in the blowpipe flame and fuses readily into a transparent bead with occluded bubbles. Colours the flame green. Almost insoluble in cold water, dissolves in hot water. Readily soluble in acids.

Genesis and Occurrence: Occurs in dried boron-bearing salt lakes, especially in salines in hot arid climate. May derive from hydroboracite orinyoite ($Ca_2B_6O_n.13H_2O$) through the weathering of salt domes containing borates. This is possible from hydroboracite, if the concentration of sodium in the percolating solutions is high enough; otherwise boronatrocalcite may be altered to inyoite. In the static zone, the mineral develops metasomatically in enclosing rocks, particularly in clays, in the process of evaporation of water from solutions.

In the USSR, occurs sporadically. In large masses, it is found in the dried salt lakes in the *Death Valley* (California); *Wales Desert* near Kern; at *Lange* (Los Angeles) with colemanite and also in the arid regions of Nevada such as *Esmeralda,* in dry regions of *Tarapaca* (Chile), and elsewhere

Uses: Like other borates, when found in quantity is a source of boric compounds.

Hydroboracite, $MgCaB_6O_n.6H_2O$

The name has no connection with boracite, from which it materially differs in composition. MgO 9.9 per cent, CaO 13.9 per cent, B_2O_3 49.5 per cent H_2O 26.7 per cent. May contain very negligible quantities of alkalis. Identified by Academician G. I. Hess in 1833.

System, monoclinic. Occurs in acicular and fibrous aggregates, sometimes beautiful spherolitic and stellate sinter formations.

Colour, Colourless or white, rarely pink, red, or grey. *Lustre, vitreous.* n_x = 1.571, n_y = 1.534, and n_z = 1.522. *Hardness,* 2. Brittle. Cleavage, {010} perfect. Specific gravity, 2.167.

Diagnostic Features: Externally indistinguishable from similar borates without optical and chemical analyses.

Fuses readily before the blowpipe flame into a bead which does not become turbid on cooling; colours the flame green. Almost insoluble in water, even on boiling. Soluble in acids on slight heating.

Genesis and Occurrence: Occurs in boron-bearing deposits of rock salt accompanied with strata of gypsum, anhydrite and clays. The clay beds are enriched in hydroboracite due to the general ability of clays to adsorb different substances, perhaps predominantly boric salts.

In the course of formation of salt domes, the hydroboracite in the zone of weathering becomes unstable under the action of percolating waters and is either re-deposited or gradually replaced by more stable borates (boronatrocalcite, inyoite, colemenite, and ascharite) formed metasomatically or deposited in leaching cavities (caverns and karsts)

Thus, considerable masses of secondary borates are formed together with gypsum as caps over salt domes.

Notable foreign localities for hydroboracite are at *Stassfurt* (Germany) and at Inyo California (U.S.A.).

Uses: Found in quantity, hydroboracite is an important material for the chemical industry as a source of boric compounds.

Colemanite, Ca_2B_6On-$5H_2O$

System, monoclinic. Crystals are short-columnar, dipyramidal in habit. More often occurs in granular spherolite masses.

Colour, white. *Lustre,* vitreous. n_x = 1.614, n_y = 1.592, and n_z = 1.586. *Hardness,* 4. *Cleavage,* {010} perfect. *Specific gravity,* 2.44. Insoluble in water. Soluble in hydrochloric acid on heating.

Large deposits of colemanite associated with gypsum and other boron minerals are found in California and Nevada (U.S.A.) in dried salt lakes and salines in districts with a hot arid climate. In Chile, near Bacos del Toro, colemanite is deposited from hot springs. In small masses it occurs at the Inner deposit.

Tandermite, $Ca_4B_10O1_9.7H_2O$

Synonym: priceite. *System*, monoclinic or triclinic. Occurs as compact white metacolloidal reniform masses with dull conchoidal fracture. n_x = 1.593, and n_z = = 1.574.

Hardness, 3.5. *Specific gravity*, 2.43. Behaviour before the blowpipe and in acids is similar to that of colemanite.

The mineral was first discovered near the port of Panderma, in the Sea of Marmora (Turkey). Occurs at *Inyo* and elsewhere in California (U.S.A.). Is found in quantity as compact masses at the *Inder* dcposit of borates (north of the Caspian Sea).

Silicates

General. The class of silicates includes a vast number of minerals—about one-third of all mineral species known in nature. Their share becomes even more important if their distribution and weight in the earth's crust are taken into consideration.

A. Y. Fersman has estimated that the silicates make up 75 per cent of the earth's crust. If we add to this another 12 per cent of free silica (mostly in the form of quartz and opal), the tremendous importance of silicium in geochemistry in general will become perfectly clear.

Many silicates are essential rock-forming minerals not only in all magmatic rocks, their hydrothermally altered varieties and contact low hardness and ready solubility in water. Can be distinguished from similar boronatrocalcite only by chemical tests (borax does not contain any calcium). Borax crystal metasomatic formations, but also in their weathering products and in many sedimentary rocks (chiefly clays and shales), and also in various crystalline schists. They are also important constituents of practically all deposits of economic minerals as accessory minerals of ores and, in a number of cases, as bearers of important metals (Ni, Zn, Be, Zr, Li, Cs, Rb, U, TR, etc.). Quite a few non-metallic economic minerals are silicates. This includes asbestos, kaolin, bleaching clays and feldspars which are materials for refractories, ceramics, etc., not to speak of building materials. A number of silicates (emerald, aquamarine, tourmaline, topaz, rhodonite, nephrite, etc.) have long been used as gemstones and ornamental stones.

The principal elements entering into the composition of silicates are:

Na, K, Li

Ca, Mg, Fe″, Mn″, Be Si, Zr, Ti

Al, Fe‴, B 0, F, H.

Numerous other elements (Rb, Cs, Ba, Sr, Pb, Zn, Ni, Co, Cu, Bi, Sb, Cr, V, Sc, Y, TR, Th, Sn, U, Nb, S, Cl, C in the form of $[CO_3]^{2-}$, and P) are present in relatively rare mineral species of the class.

In spite of the comparative fewness of principal elements (cited above) entering into the composition of silicates, an extraordinary variety of natural silicate compounds, often of highly complex constitution and variable composition, is observed in nature.

The chemical composition of oxygen salts of the classes reviewed here was expressed by rather simple formulas. Among the silicates the number of such simple compounds is relatively small. In a great number of cases chemical analysis data do not permit evolving simple formulas because of the presence of numerous impurities and certain variability of composition of the mineral species. This cannot be explained either by errors in the analyses or by the microscopic foreign inclusions in the minerals analysed. The complexity of the composition is due to the crystallochemical features of the compounds, as will be seen from the descriptions of individual mineral groups.

The constitution of silicates. The chemical nature of silicates was thoroughly studied by such eminent scientists as Vernadsky, Tcher-mak, Groth, Clarke, and many others even before the advent of the X-ray methods.

Attempts were made to solve this highly complicated

problem by careful comparison of the empirical formulas of minerals, making use of the valences of individual elements that were known at that time; by detailed investigation of the chemical and physical properties of minerals and of synthetic products obtained in the laboratory; by the study of the behaviour of minerals, and especially of their decomposition products, under natural conditions. All this resulted in the establishment of a number of empirical dependences. This was a stupendous job.

In the blowpipe flame after much swelling fuses into a transparent globule. Dissolves in water, the solution showing a slightly alkaline reaction. Taste, sweetish-alkaline, weak.

On exposure to air is readily dehydrated, becomes turbid and finally alters to a fine white powder (probably tincalconite or kernite). In the fused state is capable of dissolving metal oxides.

Genesis and Occurrence: Is formed in drying boron-bearing salt lakes. In large masses, together with other sodium salts, occurs along lake shores in Kashmir and Tibet; in some lakes in California; in the region of the Death Valley (San Bernardino) where it probably derives from the leaching of colemanite or kernite. In the USSR, borax deposits occur near mud cones in the Crimea.

Bobonatbocalcite, $NaCaB_5O_9$-$8H_2O$

Named for the elements entering into its composition. Although the choice of the name is not quite fortunate, since there is another borate of the same qualitative composition (probertite, $NaCaB_5O_9$'$5H_2O$). Synonym: ulexite.

Chemical Composition: NaO 7.7 per cent, CaO 13.8 per cent, B_2O_3 43.0 per cent, H_2O 35.5 per cent. May contain negligible K_2O and MgO and some foreign matter. System, triclinic; symmetry, pinacoidal (?). Occurs as nodules and reniform masses of asbestos-like, acicular or matted fibrous structure. Occasionally in thin platy aggregates.

Colour, white. *Lustre,* vitreous, silky. n_x = 1.519, n_y = 1.505, and n_z = 1.496. *Hardness,* 1. *Cleavage,* visible under the microscope, is in three directions crossing at right angles and thus giving the plates a rectangular habit. *Specific gravity,* 1.65.

Numerous indirect deductions served as a basis for diverse, often controversial hypothetical concepts of the constitution of many silicates of complex composition for which various structural formulas were suggested. A number of postulates advanced with amazing intuition (by Vernadsky) have been fully borne out today. Some, however, have not been corroborated by the data of the modern precise methods of research.

Inadequate as they may seem today, the early hypotheses have played their indisputable historical role in the development of the mineralogy of silicates. But for this research, especially in ascertaining the properties of the minerals, deciphering of a number of most intricate crystal structures, particularly of feldspars, would have been hampered considerably.

We shall dwell now on some terms which have been adopted in the literature on mineralogy.

On the basis of the concepts of the molecular structure of matter, the silicates were interpreted as salts (acid, normal, and basic) of numerous hypothetical acids: silicic, alumosilicic, titanozirconosilicic, and so forth. From this standpoint the silicates proper were divided into:

(1) Orthosilicates, or salts of an H_4SiO_4 acid;

(2) Metasilicates, or salts of an H_2SiO_3 acid;

(3) Pyrosilicates, or salts of an $H_6Si_2O_7$ acid, etc.

Whereas the compounds of the first type do have many properties, including the structure of the crystal lattice, characteristic of typical salts, analogous to the previously described phosphates, sulphates, etc., in the salts of the other two types, especially those rich in silica, these features are much less pronounced. Compounds of the latter type actually occupy an intermediate position between typical salts and oxides (e.g., SiO_2).

It should also be mentioned that in literature one finds other less appropriate terms for types of silicate based on the quantitative ratio between oxygen atoms of the SiO_2 ailhydride and those of the bases (derived from empirical formulas). The orthosilicates (e.g., forsterite 2MgO-SiO_2) were called-*monosilicates.*

The current concept of the constitution of silicates has developed mainly on the basis of general laws of crystal chemistry.

X-ray study of scores of silicates of widely varying composition has led to the following highly important conclusions with regard to the specific features of crystal structures of these compounds.

1. In all the silicates studied each Si^{4+} ion is always surrounded by four O^{2} ions occurring at the corners of a tetrahedron. The bonding of the oxygen ions with silicon is much stronger than with other metals acting as cations in the crystal structure of silicates. The dimensions of these silicon-oxygen tetrahedra are almost invariable. The Si-O distance is about 1.6A. This and the other X-ray data confirm that the Si —O linkage is largely covalent. Thus, the silicon-oxygen tetrahedron, i.e., $[SiO_4]^{4-}$ group, is the fundamental structural unit of all silicates.
2. The silicon-oxygen tetrahedra in the crystal structures of the silicates may be either independent SiO_4 unit cells or may be linked together in a number of wavsabo form complex anionic radicals. *The Vtetrahedra are linked only by s,* faces. Most complete linkage tetrahedron apexes are shared *earners,* never by sharing edges or viously achieved when all the four the adjacent SiO_4 tetrahedra. This is precisely the case of crystall Pictures of minerals of the quartz group with the general formula SiO_2.
3. Depending on whether the silicon-oxygen tetrahedra

are linked and on the nature of the linkage, the complex anionic radicals assume different spatial forms.

Here are the most typical examples:

(a) The complex anionic radical has the form of an *isolated* $[SiO_4]^{4-}$ tetrahedron held in the structure by cations of other metals. The total negative charge of each such group is 4 (each oxygen ion giving to silicon only half of its negative charge which is 2). This type of structure is widely represented in the orthosilicates, such as zircon, $Zr[SiO_4]$; forsterite, Mg[SiO]; garnet, $Ca_3Al_2[SiO_4]_3$, and many others.

(b) The complex anionic radical has the form of isolated $[Si_2O_7]^{6}$~ *groups* consisting of two interlinked SiO_4 silicon-oxygen tetrahedra with a common apex. The total negative charge of this group is obviously 6. The oxygen ion occurring at the common apex is electrostatically inert. Therefore, the active oxygen ions, whose residual negative charges are neutralised in the crystal structure by metallic cations, occur at the two opposite ends of the anionic radical. Compounds (pyrosilicates) with such complex anionic radicals are few, e.g., thortveitite, $Sc_2[Si_2O_7]$.

(c) The complex anion consists of three, four, and six silicon-oxygen tetrahedra sharing two apexes and thus forming closed *flat isolated rings*. In this case the complex anions will be respectively expressed as $[Si_4O_9]^{6-}$, $[Si_4O_{12}]^{12-}$, and $[Si_6O_{18}]^{12-}$.

The total valence of each such radical depends on the number of the outer oxygen ions, each having the non-compensated negative valence.

Examples are benitoite, $BaTi[Si_3O_9]$, and beryl, $Be_3Al_2[Si_6O_{18}]$. Under the older classification, the compounds of this structural type were grouped with the metasilicates, #"$[SiO_3]$, solely on the basis of its chemical formula. It should be added that N. V. Belov has demonstrated that certain silicates (milorite and tourmaline) possess a new type of anions in the form of complexes consisting of double (two-tier) rings.

(d) The complex anions constitute *unidimensional continuous chains* of linked silicon-oxygen tetrahedra. Each tetrahedron is linked to the adjacent tetrahedra by sharing two corners with inert oxygen ions, while two active oxygen ions in each tetrahedron are so arranged that one of them occurs above the Si ion (on the plane of the drawing) and the other is offset alternately upward and downward. The metallic cations are located between these *linearly extended* radicals.

In each silicon-oxygen tetrahedron, two oxygen ions completely belong to the Si ion, while the two others (the inert ones) are equally shared, as it were, by the adjacent tetrahedra. In the sum total, each Si ion holds three oxygen ions, two of which have one free valence each. Thus the composition and valence of such radicals may be expressed as follows: $n[SiO_3]^{2-}$, where *n* and indicates polymerisation. This type of the acid radical structure is characteristic of the group of pyroxenes which are classed as typical metasilicates expressed by the general formula R $[SiO_3]$.

A *double chain* or rather a band of continuously linked silicon-oxygen tetrahedra corresponding to a single chain reflected across a plane perpendicular to the drawing and parallel to the axis of the chain. Such band-type linkage of silicon-oxygen tetrahedra is characteristic of the amphiboles. It can be easily calculated that the composition and valence of such radicals within the limits of one period (5.25 A) is expressed by the formula $[Si_4O_n]^{6-}$.

It should be noted that in the silicon-oxygen tetrahedra the Si ion may be partially replaced by the Al ion which is also surrounded by four oxygen ions.

Recently, at the Research Institute of Crystallography of the USSR Academy of Sciences, N. V. Belov and other workers have established a new type of anionic radical in the form of a silicon-oxygen $[Si_3O_9]^{8-}$ chain characteristic of the minerals of the wollastonite group, $CaSiO_3$, and also in the form of double silicon-oxygen $[Si_6O_{17}]^{10-}$ chains characteristic of the OH-bearing xonotlite, $Ca_6[Si_{16}O_{17}]$ $[OH]_2$.

(e) The complex anions have the form of *two-dimensional layers* of silicon-oxygen tetrahedra. In the structure of such layered radicals, the tetrahedra are linked together by sharing three common apexes to form continuous sheets having the pattern of a hexagonal lattice. The active oxygen ions (one from each tetrahedron) are all oriented in the same direction (above or beneath the plane of the drawing) thus forming a special active layer in the tetrahedral sheet. Some of the Si ions in the silicon-oxygen tetrahedra are often replaced by Al ions with four-fold coordination. Each such sheet is in some way linked to other structurally identical sheets by a layer of active oxygen ions and through metallic cations. Examples of such crystal structures are found in lamellar minerals possessing excellent unidirectional cleavage: micas, talc, chlorites, etc.

(f) The complex anions are comprised of *three-dimensional frameworks* of silicon-oxygen tetrahedra, in which each oxygen ion is shared by two adjacent tetrahedra. There is not a single tetrahedral corner with an active oxygen ion in it. Such frameworks are found in minerals of the quartz group, which constitute, as it were, pure anhydrite with the SiO_2 formula. It is true that in such cases, some of the Si^{4+} ions are always replaced by Al^{3+} ions with the same coordination number. The general chemical formula of complex anions with a framework structure may be written as $[(Si_{n\ X}Al_X\ O_{2n}]\sim$. Because of the partial substitution of Al^{3+} for the Si^{4+} ions (the total number of oxygen ions Aluminium-silicon-oxygen framework in crystal structure of nosean remaining the same), the radical has a certain residual negative charge, as can be easily calculated. Examples are feldspars Na [SiaAlO], $Ca[Si_2Al_2O_8]$, and many other minerals. The Na, Ca, and other cations which balance the residual negative valence of the anionic frameworks are enclosed within the frameworks (in the respective "voids" of the structure).

Thus, the general features of the crystallochemistry of silicates outlined on the basis of the division of complex anions of the SiO_4 tetrahedra into 0-, 1-, 2-, and 3-dimensional radicals.

Types of Silicon-oxygen Anions on the Basis of Si:O Ratios

Before the advent of X-ray methods, some authorities maintained that the Al ions, by analogy with Mg, K, Na, and other metals, played the role of bases in the silicates, whereas V. I. Vernadsky argued that the Al of the silicates entered into the composition of the acid radicals along with Si and, therefore, he deduced numerous alumosilicic acids, their anhydrites and salts (alumosilicates).

The problem was solved by means of X-ray examination of the crystal structures of the silicates. It has been demonstrated that *alumosilicates* do exist among the silicates. In many cases, however, the existence of silicates of aluminium was also established in a large number of cases, often together with other metals.

Thus, Al^{3+} plays a dual role in the constitution of the silicates, i.e., either as a constituent of the anionic radicals, surrounded, like Si^{4+}, by four oxygen ions; or as an independent cation which, alone or together with other metal cations, neutralises the negative anionic charge, and is surrounded, like Mg^{2+}, by six oxygen or hydroxyl ions.

In quite a few cases, in one and the same silicate some Al ions occur in the anionic radical and some in the cations which occupy the spaces between the negatively charged radicals.

In the light of these discoveries, many hitherto puzzling aspects of the problem have found a rational explanation on the basis of most precise physical data, in full conformity with the properties of the minerals.

Thus in the alumosilicates the oxygen ions around Al^{3+} have the same four-fold coordination as around Si^{4+}. The possibility of the silicon-oxygen tetrahedra being replaced by alumooxygen tetrahedra may be inferred on a purely geometrical basis: the ratio of the O^{2-} and Al^{3+} ionic radii is 0.43, i.e., comes close to the boundary between the possible four-fold coordination as for Si^{4+}, and six-fold coordination as for Mg^{2+}. However, the replacement of Si^{4+} by Al^{3+} is fraught with important consequences that essentially affect the constitution of alumosilicates.

Indeed, whereas the SiO_4 silicon-oxygen tetrahedron has four unsaturated valences, the AlO_4 alumo-oxygen tetrahedron has five (8 — 3 = 5). Hence, each alumo-oxygen tetrahedron in the complex anionic radical increases its negative charge by a unit which has to be balanced by some positively charged cation in accordance with the stereometric structure of the given compound. This can be easily demonstrated on the example of compounds possessing crystal structures of the framework type.

Quartz, Si_4O_8

Albite, Na [SigAlOg]

Nepheline, Na_2 $[Si_2Al_2O_8]$

Anorthite, $Ca[Si_2Al_2O_8]$, etc.

The Si, Al ratio need not to be an integer, but the equilibrium of negative and positive charges is absolutely essential for the stability of a crystal structure. This is why the data of chemical analyses of the alumosilicates do not always simply correspond to the chemical formula of the mineral analysed.

It should be noted that the formation of the AlO_4 group is possible under certain definite conditions. Frameworks composed of tetra hedral $A1D_4$ and SiO_4 groups may be formed when the lattice structure contains relatively large cations with a low charge (Na^1, K^1, Ca^2, Ba^{2+}, etc.). Such are the numerous members of the groups of feldspars and zeolites. Characteristically, when alkaline and alkaline-earth metals are leached, the AlO_4 groups collapse, and in the kaolinite, which results from the decomposition of feldspars, Al^{3+} is already in six-fold coordination. The same is true of other types of crystal structure whose anionic radicals include an AlO_4 component.

In an alumina-rich environment, AlO_4 groups are formed at high temperatures. It has been shown experimentally that such minerals as kaolinite and disthene, all of whose Al^{3+} is always in sixfold coordination, partly alter under such conditions (1400 to 1500°) to silli-manite or mullite, i.e., to silicates which also contain AlO_4 groups with four-fold coordination. It should be noted that their heating curves show pronounced exothermal effects, fully conforming to the point of rearrangement of the crystal structure.

In this connection, mention should be made of the so-called kaolin nucleus. Observing the natural breakdown of minerals and studying experimental research data, V. I. Vernadsky came to the conclusion that alumosilicates, feldspars in particular, should contain a specific $Al_2Si_2O_7$ nucleus, highly stable in natural environment. The very idea of a close tie between Al and Si in alumina-bearing silicates was of invaluable assistance in the subsequent X-ray deciphering of the extremely complex structures of the alumosilicate minerals. It has been found, however, that by far not all the alumina-bearing silicates were alumosilicates, i.e., compounds in which Al plays the same part as Si. Even the alumosilicates proper, with their continuous spatial extension of alumo-silica-oxygen radicals, strictly speaking, contain no detached or isolated groups like the hypothetical "kaolin nucleus". Kaolinite itself, a product of weathering of the alumosilicates, in effect proved to be a silicate of aluminium, not an alumosilicate or "alumosilicic acid".

Many silicates contain so-called additional anions neutralising the excess positive charge of the cations.

Presumably, in some cases, the $[OH]^{1-}$ and F^1 may replace the oxygen ions within the complex anionic radicals; usually, however, such replacement takes place when the O ions do not enter into the composition of the silicon-oxygen tetrahedra.

Finally, a number of silicates contain H_2O, mostly of zeolitic nature. The H_2O molecules are usually very weakly held in the voids or channels of crystal structures.

In all the silicates the number of oxygen ions exceeds that of ions of the other elements. Since the oxygen ions have considerably longer ionic radii than the cations, it is only natural that the size of unit cells depends primarily, as in other oxygen compounds, on the number of oxygen ions per cell and on their spatial positions.

Hence, silicates that are the members of the same isomorphous series should have the same number of oxygen ions.

Phenomena of heterovalent isomorphism along with isovalent isomorphism are widespread in the minerals of the silicate group.

A classical example of heterovalent isomorphism is the plagioclase series: $Na[Si_3AlO_8]$—$Ca[Si_2Al_2O_8]$. Here Na^{1+} is replaced by Ca^{2+} of about the same volume. The accompanying increase of the positive charge by a unit is balanced by a corresponding substitution in the radical: one Si^{4+} ion being replaced by an Al^{3+} ion, or one $[SiO_4]^{4-}$ anion by an $[AlO_4]^{5-}$ anion, i.e., the negative charge is also increased by a unit.

Thus, the above and many other similar instances strictly bear out the conclusion that the components of isomorphous mixtures must have: (a) the same number of oxygen ions, since the size of the unit cells changes little; and (b) the same total valence of the replaced and the replacing ions this being dictated by the requirement that the positive and negative charges in a crystal structure be equal.

As far as the quantitative ratio between the substituting and substituted ions is concerned, their numbers are, as a rule, equal too. It will be shown later, however, in *descriptions of* individual minerals, that there exist fully justified exceptions to this. For instance, $3Mg^{2+}$ are replaced by $2AP^{+}$, $3Fe^{2+}$ by $2Fe^{3+}$, etc., the total cationic valence remaining the same (there are no changes whatever in the anionic radical). It should be emphasised that such substitution is confined to definite types of crystal structure, the most favourable conditions being found in the sheet structures.

As in other classes, among the silicates there are instances of restricted miscibility of mineral species, as well as phenomena of disintegration of solid solutions.

The restricted miscibility of isostructural, or structurally similar, compounds is due to the considerable difference in size of the substituting and substituted ions. Thus, the silicates of Ca and Mg, which are of the same type, have long been known to yield binary compounds (monticcllite, diopside, etc.) which have restricted miscibility with their components (end members of the series), and even that at high temperatures only.

Only between $Mg_2[SiO_3]_2$ and $MgCa[SiO_3]_2$ (clinoenstatite—diopside) there exist continuous isomorphous mixtures formed in natural conditions at high temperatures.

Highly important structural characteristics of the silicates are the coordination numbers, since they control isomorphous substitutions. Some of the cations have two or even three coordination numbers. Listed below are the elements commonly occurring in the silicates, and their coordination numbers established for crystal structures that have been studied to-date:

Si—4 Fe— 4 and 6 Na — 6 and 8 Zr — 6 and 8

Be—4 Mg—6 and 8 Ca—6,7 and 8 K— 6 and 10

Zn — 4 Fe — 6 and 8 Li—6 Ba—12

As for Fe^{3+}, this element in a number of cases enters into the complex anion as a substitute for Si^{4+}, similarly to Al^{3+}, and with exactly the same effects. Such substitution has been proved for a number of strongly ferruginous silicates.

Classification of Silicates: At present, with the great progress in X-ray methods of investigation of minerals, the silicates naturally must be classified not only on the basis of their chemical composition, but also according to the types of crystal structures.

Different methods of linkage of the silicon-oxygen tetrahedra leave a direct impress both on the chemical formulas of the minerals and on the morphological features of the crystals, as well as on their physical properties (refractive index, birefringence, specific gravity, and so forth).

The types of structures may be classified in different orders. Some authors classify silicates in the order we have adopted in the above review: starting with structures with independent silicon-oxygen tetrahedra and ending with continuous three-dimensional framework. Others adopt the reverse order, beginning with quartz which has a typical framework crystal structure.

We shall adhere to the former classification, i.e., start with those groups of silicates which correspond to typical salts and are the simplest and closest in constitution to the minerals reviewed earlier in the oxygen-salt class. As different from the latter, we shall divide the class of silicates into subclasses, as follows:

Sub-class A. Silicates with isolated SiO_4 tetrahedra in crystal structures.

Sub-class B. Silicates with isolated groups of SiO_4 tetrahedra in crystal structures.

Type 1: With isolated Si_2O_7 groups.

Sub-class C. Silicates with continuous chains of SiO_4 tetrahedra in crystal structures.

Type 1: With single chains.

Type 2: With double chains.

Sub-class D. Silicates with continuous sheets of SiO_4 tetrahedra in crystal structures.

Sub-class E. Silicates with continuous three-dimensional frameworks of (Si, Al)O_4 tetrahedra in crystal structures.

Silicates with Isolated SiO_4 Tetrahedra in Crystal Structures

The salient feature of crystal structures of the silicates in this sub-class is the occurrence of their unit cells as detached tetrahedral $[SiO]^{4-}$ anions. These tetrahedra stand isolated in the crystal structure, i.e., none of the oxygen ions surrounding the Si ion are shared by the adjacent silicon-oxygen tetrahedra.

From the chemical standpoint, these silicates were regarded as orthosilicates, i.e., as salts of a hypothetical H_4SiO_4 acid. The most important cations in the silicates of this type are Mg^2, Fe^{2+}, Ca^{2+}, partly Ni^{2+}, Co^{2+}, Mn^{2+}, Zn^{2+}, and also Al^{3+}, Fe^{3+}, partly Mn^{3+}, Cr^{3+}, sometimes Pb^{2+}, Be^{2+}, Ti^{4+}, Zr^{4+},Th^{4+}, and occasionally Nb^{5+}. Na^{1+} and K^{1+} alkalis occur as a rare exception. Rare earths may also take part in the structure of crystal lattices, along with calcium, sodium, and partly thorium. Aluminium, in distinction from the other types of silicates, is never present in crystal structure as complex anions with four-fold coordination, i.e., does not replace Si in its tetrahedral groups. Presumably phosphorus alone is capable of replacing silicon ions. The physical properties of silicates of this type are rather peculiar and are determined by the specific features of the compact crystal lattices. Crystal forms are usually isometric. The minerals possess high hardness and relatively high specific gravity, owing to the close packing of ions. This also explains their relatively high refractive indices. Intense colouring is characteristic only of the varieties containing chromophores.

Zircon Group

Included in this group are the orthosilicates of tetravalent Zr and Th, crystallising in the tetragonal system.

Zircon, $ZrSiO_4$

From corrupted Persian "tsar" meaning gold and "gun" colour. Synonym: hyacinth. Often present as an accessory mineral (in minor quantities) in acid and alkaline igneous rocks (granites, syenites, and nepheline-syenites).

Chemical Composition: According to formula: ZrO_2 67.1 per cent; (Zr 49.5 per cent), SiO_2 32.9 per cent. Practically always has a slight admixture of Fe_2O_3 (up to 0.35 per cent and more), often CaO (0.05 to 4 per cent), and sometimes Al_2O_3. Always contains hafnium oxide (sometimes up to 4 per cent of HfO_2) and in *alvite* from Kragero (Norway) even 16 per cent. May contain Y_2O_3 and rare earths, chiefly Ce_2O_3 (hagatalite) sometimes up to 16 per cent with P_2O_5 content of 4 to 5 per cent (amagutilite). Certain varieties may contain Nb and Ta (naegite), ThO_2, up to 7 per cent and even 12 per cent (hogtveitite). Occasionally contains negligible Sn and Be (in alvite, the content of BeO + Al_2O_3 may reach 15 per cent), etc.

Finally, varieties are known containing much P_2O_5 (oyamalite). *Mala-cons* and cyrtolites are rich in radioactive substances.

System, tetragonal; *symmetry,* ditetragonal dipyramidal. Mostly in crystals, more seldom in irregular grains.

Crystal Structure

X-ray investigation shows a typical radical-ionic structure comprised of anionic SiO_4 groups and Zr^{4+} cations surrounded by eight oxygen ions. The SiO_4 etrahedra alternate parallel to L^4 with the Zr ions. Thus, zircon sharply differs in structure from rutile whose crystals externally resemble very closely those of zircon, but whose structure is analogous to that of anhydrite, $Ca[SO_4]$.

Habit: Short-columnar, isometric, less often dipyramidal. Common forms: tetragonal {100} and {110} and a tetragonal dipyramid (111, 276). Twins are geniculate, like those of rutile, but occur far less often.

Colour: Colourless, but mostly yellow, orango, red, and often green. Lustre, adamantine, sometimes greasy. Malacons are usually dark brown. n_x = 1.968 to 2.015 and n_y = 1.923 to 1.980. *Hardness*, 7 to 8. For varieties that underwent metamict decay it diminishes to 6; moreover, they are remarkably tenacious while common zircons are brittle. Cleavage, relatively rare parallel to {110}. Fracture uneven; for altered varieties, conchoidal. Specific gravity, 4.68 to 4.70; for altered varieties such as cyrtolites declines to 4.7 and even 3.8 (alteration causes increase of volume and slight distortion of the external crystal form). Other properties. Malacons and cyrtolites are, as a rule, radioactive.

Diagnostic Features: Zircon crystals have characteristic tetragonal short-columnar and less often dipyramidal habit. May be mistaken for: (1) rutile (distinguished by hardness and refractive indices), (2) cassiterite (distinguished by specific gravity, paragenesis, chemical reactions, and lower birefringence under the microscope), (3) thorite (distinguished by hardness, chemical reactions, and much higher birefringence), and (4) monazite which occurs in analogous conditions (distinguished by hardness and by habit, monaaite usually being tabular).

Infusible before the blowpipe- Insoluble in acids. Decomposes when fused in a powdered state with soda; the flux, when dissolved in dilute HCl, colours the lithmus paper orange (reaction for Zr).

Genesis and Occurrence: Zircon occurs as small, rare disseminated crystals in magmatic intrusive rocks; nepheline-syenites, granites, diorites, gneisses, but more often and in larger crystals in syenitic and granitic *pegmatites*. In metamorphosed sedimentary rocks (crystalline schists and paragneisses) it is found as rounded relict grains.

Because of the idiomorphism of zircon crystals with regard to all the accompanying minerals, it is thought to be one of the first minerals to crystallise from magma. This concept does not agree, however, with a number of facts observed in nature: (1) zircon is rather often closely associated with later minerals containing volatiles (apatite, fluorite, micas, etc.); (2) according to Lacroix, in the Madagascar nephelines where its content may reach 7 per cent, zircon often occurs in cracks in partly solidified parent rock; (3) in effusive rocks (trachytes, basalts) it occurs as tiny crystals in druse-lined cavities; (4) it occurs pseudomorphous after minerals of magmatic origin such as alkaline hornblende (riebeckito), and others. The idiomorphism of zircon in many instances may be duo to the fact that, despite its relatively late formation, similar to metacrystals (pyrite, magnetite, apatite, etc.), it is capable of forming perfect crystals.

As a chemically inert mineral, zircon is easily liberated from its accessories in the course of weathering and passes into placers, and hence as rounded grains, into sedimentary rocks.

Zircon predominantly occurs in miaskite and syenite pegmatites together with black mica, nepheline, albite, apatite as well as with various minerals containing TR, Nb, Ta, Th and U (pyrochlore, aschinite, samarskite, etc.). Crystals are usually not large, up to 1 cm across, though occasionally large individuals are found.

The largest specimen ever found (in 1837), about 3.5 kg in weight, consisted of several individuals. A study was made of oriented intergrowths of zircon with pyrochlore along (111) planes, the face of the pyrochlore octahedron coinciding with that of the zircon tetragonal dipyramid. It transpired that the ordered nature of this intergrowth was explained by the isometricity of the [110] and [112] edges of zircon and pyrochlore respectively. It should be rioted that when the intergrowths were broken, no idiomorphism was observed in minerals in relation to each other: their fractured composite planes were uneven and irregular. This suggests a simultaneous development of both.

3

Sulphides and Sulphosalts

The sulphur, selenium, tellurium, arsenic and antimony compounds of metals or at least many of them are minerals of great economic importance and are an essential part of numerous deposits of metallic ores. The vast majority of these minerals are sulphur compounds (sulphides and sulphosalts). All of them, with the exception of hydrogen sulphide, occur in nature as solids. In all, 40 chemical elements in some form or another combine with sulphur. The most important of them are H, V, Mn, Fe, Ni, Co, Cu, Zn, (Ca), Ce, As, Mo, Ru, (Rh), (Pd), Ag, Cd, (In), Sn, Sb, (Re), (Os), (Ir), Pt, Hg, Tl, Pb, and Bi. The elements given in parentheses do not combine with sulphur and occur only as isomorphous admixtures.

Sulphur compounds, according to V. I. Vernadsky, account for no more than approximately 0.15 per cent of the earth's crust by weight, Fe being the predominant metal constituent in these compounds. The sulphur compounds of all other elements, excluding hydrogen sulphide, constitute a negligible percentage of the weight of the earth's crust (about 0.001). The elements which form the most common sulphur compounds are Zn, Pb, Cu, Ag, Sb, Bi, Ni, Co, Mo, and Hg. ***Selenium*** compounds (selenides) are known for the following elements: H, Cu, Ag, Hg, Pb, and Bi. Moreover, selenium is often present in sulphur compounds as an isomorphous admixture.

Tellurium compounds (tellurides) are more common in nature than selenides, although the elements which combine with tellurium are fewer in number, namely, Cu, Ag, Au, Hg, Pb, Bi, Ni, and Pt. All these elements form separate minerals.

Simple *arsenic* compounds (arsenides) are known for only a few elements, namely, Fe, Ni, Co, and Pt. Complex compounds in the form of sulphosalts are more widespread as sulphoarsenites, chiefly of Cu, Ag, and Pb. Simple compounds of *antimony* (antimonides) are known only for Ni, whereas complex compounds such as sulphoantimonites of Cu, Ag, and Pb, are rather common.

Simple *bismuth* compounds are not known in nature, but sulpho-bismuthites of Cu, Ag, and Pb have been found.

Selenium, tellurium, arsenic, antimony and bismuth compounds are incomparably fewer than sulphur compounds.

Chemical features of compounds. The periodic table shows that sulphur compounds are typical of the elements almost exclusively located on the right-hand side of the table, especially of the metals of:

H^{1}											
H^{2}	Li^{3}	Be^{4}	B^{5}	c^{6}	N^{7}	O^{8}					F^{9}
$N.^{10}$	Na^{11}	Mg^{12}	Al^{13}	Si^{14}	P^{15}	s^{16}					Cl^{17}
Ar^{18}	K^{19}	Ca^{20}	Sc^{21}	Tl^{22}	v^{23}	Cr^{24}	Mu^{25}	Fe^{26}	Co^{27}	Ni^{28}	Ci^{29}
Zn^{30}	Ga^{31}	Ge^{32}	As^{33}	Se^{34}	Br^{35}						
Kr^{36}	Rb^{37}	Sr^{38}	Y^{39}	Zr^{40}	Nb^{41}	Mo^{42}	To^{43}	Ru^{44}	Rh^{45}	Pd^{46}	Ag^{47}
Cd^{48}	In^{49}	Sn^{50}	Sb^{51}	Te^{52}	I^{53}						
Xe^{54}	Cs^{55}	Ba^{56}	TR^{57-71*}	Hf^{72}	Ta^{73}	w^{74}	Re^{75}	Os^{76}	Ir^{77}	Pt^{78}	Au^{79}
Hg^{80}	Tl^{81}	Pb^{82}	Bi^{83}	Po^{84}	At^{85}						
Rn^{86}	Fr^{87}	Ra^{88}	$Ac^{89-103*}$	Th^{90}	Pa^{91}	U^{92}					

Elements characteristically forming sulphides and similar compounds

A place apart among the elements on the left-hand side of the table held by hydrogen whose sulphide, H_2S, is analogue of water and play tremendous role in the formation of sulphides.

The pronounced affinity of sulphur with heavy metals is shown by the fact that it reacts with them to form compounds which are almost insoluble in water. In contrast to these, the corresponding compounds of light metals (Na, K, Mg, Ca, etc.) are found only in aqueous solutions.

Chemically, sulphides and sulphosalts under consideration are similar to simple oxides and oxygen salts and, indeed, the formulas of these two types of compounds have much in common (e.g., Cu_2O and Cu_2S, HgO and HgS, Sb_2O_3 and Sb_2S_3 Cu_3AsS_4 and $Cu_3[AsO_4]_2$, etc.).

The majority of simple sulphur, selenium and tellurium compounds, from the chemical viewpoint, derive from H_2S, H_2Se, and H_2Te (e.g., XS, ZnS, HgSe, PbS, PbTe, Sb_2S_3, etc.).

In complex sulphur compounds, these simple sulphides either form *rinary compounds* or such which chemically are regarded as a separate Jass of *sulphosalts.*

Furthermore, there are simple compounds of the AX_2 type or *di-wlphides* which, according to V.I. Vernadsky, are derived from H_2S_2. Comparing them with oxides, Vernadsky considered them to be analogues of peroxides. Indeed, on heating, disulphides yield some of their sulphur content in the same way that peroxides lose some of their oxygen.

Arsenides and *antimonides* (simple compounds of metals with arsenic and antimony) materially differ in chemical nature from sulphides. Contrary to the older conception, neither arsenic nor antimony are capable of isomorphous replacement of sulphur. Examples of arsenides and antimonides are NiAs, NiSb, $FeAs_2$, $CoAs_2$, etc. Close to disulphides and diarsenides, both chemically and physically, are *sulphoarsenides* (FeAsS, etc.) and *sulphoantimonides* (NiSbS, etc.).

Many compounds of this class form solid solutions with each other, either as continuous series or with limited miscibility in the solid state. Such are, for instance, the HgS—HgSe and $CoAs_2$—$NiAs_2$ — $FeAs_2$ series. Finally, there are numerous polymorphous modifications.

Crystal structure and physical properties: According to X-ray data, sulphides and similar compounds should be classed with ionic compounds, although in their majority in a whole number of characteristic properties, they differ sharply from typical ionic oxygen compounds and are closer to native elements than to oxides or oxygen salts. These differences are due to the inherent properties of the atoms or ions entering into the composition of sulphides, selenides, tellurides, arsenides, and antimonides.

As compared with oxygen, the ions of S, Se, Te, As, and Sb have much greater radii, are more easily polarised and form weak homopolar bonds. On the other hand, the metallic ions which form compounds with them and which are situated on the right-hand side of the periodic table (in the iron family and in the sub-groups of the long periods) have an 18-electron outer shell and are strongly polarising. Complex ions, though present in complex sulphur compounds, such as sulphosalts, have bonds which are not as strong as those of silicates and other oxygen salts.

Polarisation phenomena in the crystal structures result in pronounced merging of many electrons of oppositely charged neighbouring ions, which may be inferred from the pronounced *metallic lustre* characteristic of sulphides and similar compounds, and generally typical of metals. This agrees well with the fact that most sulphur, arsenic and other similar compounds are *good conductors of electricity* and also explains why there are fewer metal atoms as compared with the number of metalloid atoms (for instance, in pyrrhotite, FeS), and the frequent absence of strict Dalton ratios in minerals.

In S, Se and Te compounds, metallic properties become more pronounced as the sulphur is replaced by selenium and tellurium. In the iron group these properties increase from Mn to Ni. The same is true in the As, Sb, and Bi series where metallic properties increase towards the bismuth end. In fact, bismuthite (Bi_2S_3) has a stronger metallic lustre than antimonite (Sb_2S_3), whereas orpiment (As_2S_3) is translucent and has an adamantine or sub-metallic lustre.

Mode of occurrence. The overwhelming majority of accumulations of sulphides and similar compounds occur in ore deposits of hydrothermal origin. This warrants the assumption that heavy metals are carried out of the magmatic hearth as volatile or mobile compounds and precipitate at lower temperatures and pressures mainly in the form of sulphur compounds. On the basis of experiments, it is considered possible that these compounds could have been transported by hydrothermal solutions as true solutions of the heavy metals in combination with Cl, F, and B, from which sulphur compounds were precipitated as poorly soluble products of exchange decomposition (A.G. Betekhtin) at lower temperatures, due to increased dissociation of dissolved H_2S into H+ and S^{2}~ ions. The solutions may have also contained complex salts with sulphur alkalis or sulphohydrates (NaHS, KHS, etc.) which readily dissolve even gold. Their decomposition at lower temperatures may have yielded native gold along with common metal sulphides (gold sulphides do not occur in nature, whereas gold tellurides do).

In sedimentary argillaceous rocks and bitumen- and coal-bearing seams, sulphides are formed in a different way. The most common sulphides in these rocks are pyrite and marcasite (FeS_2). These are formed in reducing conditions, in the presence of H_2S which is a decomposition product of protein substances in organic remains, either in the complete absence of oxygen or when it is deficient. It is also probable that not infrequently bacteria take part in this process.

In the course of weathering, i.e., in the presence of water and oxygen, practically all these minerals readily oxidise and decompose, yielding first mostly sulphates which are easily soluble in water, and then hydroxides, oxides, carbonates, and other oxygen compounds usually found in the so-called zone of oxidation of ore deposits. The only exceptions are certain minerals, such as cinnabar (HgS), sperrylite and laurite (RuS_2), which are chemically stable under these conditions.

Classification of sulphides and similar compounds. On the basis of the chemical characteristics of individual types of compounds, they fall into two primary classes.

Class 1: Simple and binary sulphides, selenides, tellurides, arsenides, and antimonides.

Class 2: Sulphosalts, i.e., minerals which are very similar in chemical constitution to salt-like compounds.

Simple Sulphides and Similar Compounds

With the exception of hydrogen sulphide, all sulphides and similar compounds occur in the earth's crust as solid crystalline substances. Depending on the type of chemical compounds, they are further divided into different principal groups of minerals (sulphides, arsenides, and antimonides).

Chalcocite Group

This group comprises copper and silver minerals of the type -4_2Se, and -A_2Te. Many of them exist in nature in two modifications: a higher-temperature cubic modification and a lower-temperature orthorhombic or monoclinic modification.

Chalcocite, Cu_2S: The name is derived from the Greek "chalcos" meaning copper. Synonym: copper glance.

Cu_2S occurs in three modifications: a low-temperature orthorhombic modification stable below 91° (chalcocite proper or ^-chalcocite); two high-temperature modifications (over 91°) which are respectively hexagonal and cubic (a-chalcocite). The hexagonal modification, the composition of which strictly conforms to the formula Cu_2S, is unstable and on decomposition alters to the cubic modification, a-chalcocite (digenite), which has an antifluorite structure and a composition, expressed by the formula $Cu_{2-x}S$, where x = 0.03 to 0.11 (a_0 = 5.55). The prerequisite for the stability of a-chalcocite is a statistic blank of about 10 per cent of Cu positions, with the replacement of another 10 per cent of univalent copper by divalent copper.

The high-temperature hexagonal modification of Cu_2S has a structure of closest hexagonal packing. The copper ions are located at the centres of all the triangles comprised of sulphur anions of each layer of packing, a_0 = 3.89, c_0 = 6.68.

Natural chalcocite is often a mixture of low-temperature orthorhombic (chalcocite) and a-chalcocite.

Chemical composition: Cu 79.9 per cent, S 20.1 per cent. Usually contains as impurities Ag, sometimes Fe, Co, Ni, As and Au. Some of them, at least the latter, are mechanical impurities.

System, orthorhombic; symmetry class, rhombic-dipyramidal. Occasionally pseudocubic. The following faces are characteristic of chalcocite which crystallises at temperatures under 91°: prismatic {110}, {021}, and {023}; pinacoidal {001}; dipyramidal {111}, {112}, {113}, etc. The crystal structure of orthorhombic chalcocite is highly complex and has not yet been studied in detail. Habit. Crystals are rare. Most often they occur as thick tablets parallel to {001} and short columns parallel to the *a* axis 89. They are often hexagonal 90, owing to formation of trillings with {110} as the composition planes. Penetration twins are also sometimes observed with (032) as the twinning plane, less often (112) aggregates. Usually occurs as finegrained massive or as impregnations in pseudomorphs after bornite and chalcopyrite, sometimes after sphalerite, galena, covellite, pyrite, etc.

Colour: lead grey streak, dark grey *Lustre,* metallic. *Hardness,* 2 to 3. Poorly malleable. *Cleavage,* {110} indistinct. *Specific gravity,* 5.5 to 5.8. Good conductor of electricity.

Diagnostic Features: Lead-grey colour, low hardness, low malleability (the point of a knife leaves a shiny scratch, which distinguishes it from otherwise similar tetrahedrite). Green colour of solution in HNO_3. Typical associates with copper minerals, most often with bornite.

Fusible in the blowpipe flame, colouring the flame blue. After roasting on charcoal with soda gives a globule of copper. Dissolves in HNO_3 more readily than in other acids with the separation of sulphur.

Genesis and Occurrence: Natural orthorhombic chalcocite is formed both endogenously and exogenously, but only at temperatures below 1°C.

As an endogenous mineral, it is occasionally found in certain *tydrothermal* copper-rich and sulphur-poor sulphide deposits. In this phase, its most frequent paragenetic associate is endogenous bornite. Examples of such deposits are Jezkazgan (Kazakhstan) and Redruth in Cornwall (Britain).

However, the bulk of chalcocite is formed *exogenously* in the so-called zones of secondary sulphide enrichment in all copper-sulphide deposits. Similarly to other secondary copper sulphides, chal-cocite is formed by reactions between primary sulphides and copper sulphate solutions which percolate from the zone of oxidation of copper deposits. Often, it develops metasomatically in place of secondary born-te. Sometimes it directly replaces primary chalcopyrite, and occasionally galena, sphalerite, and other sulphides of primary ores.

Gases are known of chalcocite being formed from cupriferous solutions in *sedimentary* rocks containing organic remains in the form of pseudomorphs, which retain all the structural details of the remains, principally wood.

In the zone of oxygen weathering chalcocite is unstable and upon decomposition alters to cuprite (Cu_2O), malachite, azurite, and other oxygen compounds. Incomplete oxidation of chalcocite often results in he formation of native copper after the reaction:

> Large deposits of chalcocite ores are comparatively rare. In considerable quantities, the ores are formed in the lower parts of oxidation zones, if copper-rich sulphide deposits, the upper parts of which had been

thoroughly reworked by surface agents. Ghalcocite, which in this case S, the principal copper mineral, forms zones of secondary sulphide enrichment

In the USSR, massive chalcocite ores had been mined at *Turya* in the Northern Urals, from which well-formed crystals have also been recorded. They were studied in detail by Academician P.V. Yeremeyev, Huge deposits of poor disseminated chalcocite ores are found in the Soviet Union at *Kounrad,* Kazakhstan (north of Lake Balkhash), at *Al-malyk* (south of Tashkent), and elsewhere.

Well known deposits occur in the *Butte,* Montana (U.S.A.), where chalcocite in paragenesis with bornite, enargite, pyrite and other minerals is found in primary ores well below the ground-water table, i.e., as an endogenous mineral.

Uses: Ghalcocite is the most copper-rich sulphide which makes chalcocite copper ore the most important sulphide ore of industry. A very large part of the world's production of copper is obtained from chalcocite ores. Such, in particular, are the huge deposits of disseminated ores of the Kounrad deposit, the mining of which is economically worthwhile, despite the relatively low copper content.

Argentite: Ag_2S. The name comes from the Latin "argentum" meaning silver. Synonym: silver glance. "Silver black" is a pulverulent variety of silver sulphide and occurs in association with compact argentite.

Ag_2S occurs in two modifications: (1) high-temperature cubic which is stable at temperatures above 179°C and is known as *argentite,* and (2) a low-temperature orthorhombic modification which forms at temperatures below 179°C and is called *acanthite.* At temperatures below 179°C the cubic modification undergoes paramorphic alteration to the orthorhombic, as shown by X-ray examination of the cubic crystals.

Although special names have been suggested for each

modification of silver sulphide, the general name of "argentite" has remained in use in mineralogy and is applied also to low-temperature paramorphs of the high-temperature modification.

Chemical Composition: Ag 87.1 per cent; S 12.9 per cent. Guis often found among isomorphous admixtures in argentite. Usually is also contaminated with compounds of Pb, Fe, Sb, etc.

System, cubic; symmetry, hexoctahedral. Low-temperature argentite (acanthite) is mono-clinic. Occurs as imperfect crystals (paramorphs): cubes, cubic octahedra 91, occasionally rhomtite crystal bododecahedra. (common form).

Colour: Lead grey. Lustre on fracture, metallic. *Hardness,* 2 to 2.5. Malleable. *Cleavage,* {110} and {100} imperfect. *Specific gravity,* 7.2 to 7.4. Other properties, conducts electricity only at high temperatures. A polished surface tarnishes within a few seconds on exposure to strong light.

Diagnostic Features: Difficult to recognise with the naked eye. Often associated with "silver black", sometimes with native silver.

After roasting on charcoal in the blowpipe flame gives a malleable silver globule. Dissolves in nitric acid with the separation of sulphur; when hydrochloric acid is added, a thick curdy white precipitate of AgCl is formed, which is soluble in ammonia.

Genesis and occurrence: Argentite occurs in hydrothermal sulphide silver-ore deposits, often in paragenesis with native silver and other argentiferous minerals.

However, this mineral, or acanthite to be more precise, most frequently occurs in the lower parts of *oxidation zones* of argentiferous sulphide ore deposits, in association with cerussite ($PbCO_3$), cerargyrite (AgCl), native silver, etc. Pseudomorphs of argentite have been found after native silver and many minerals of complex composition— sulphides, arsenides, and antimonides of silver (proustite, pyrargyrite, stephanite, etc.).

Argentite is rarely found in large accumulations important

in their own right. Considerable masses have been found together with native silver, at the *Kongsberg* deposit (Norway), and also at Zacatecas and Guanajuato and other deposits in *Mexico*, where it occurs in association with sulphosalts of silver (polybasite, pyrargyrite, proustite, etc.).

In the USSR, it occurs as selvages, less often as small masses, at *Zmeinogorsk* (Altai), Nerchinsk (Eastern Trans-Baikal Region), and Verkhoyansk (Eastern Siberia).

Uses: As an associate of other argentiferous minerals, argentite is a source of silver, since a minimum content of 0.02 per cent of silver makes mining worthwhile.

Galena Group

Of the minerals of this group we shall discuss only galena, the one which is most widespread in nature.

Galena, PbS: From Latin "galena" meaning lead ore. Synonym: lead glance. Variety: *selenium galena.* There is a physical variety called "bleischweif" which is compact dull close-grained mass.

Chemical composition: Pb 86.6 per cent, S 13.4 per cent. The most frequent impurities are Ag (up to tenths of a per cent), Cu, Zn, sometimes Se (selenium galena), Bi, Fe, As, Sb, Mo, rarely Mn, U, etc. In the majority of cases these elements are present as microinclusions of foreign minerals.

System, cubic; symmetry, hexoctahedral. The crystal structure, in which the sulphides of the galena group crystallise, is of the halite type. The sulphur anions are spaced according to the law of cubic (treble-layer) closest, packing, while the lead cations fill all the octahedral spaces between the anions. The unit cell is fundamentally a cubic face-centred structure in which the ions are located at the corners of the cube and face centres of the cell, with the difference that two types of ions take part in the structure. If the unit cell is divided into smaller cubes, their corners will be alternately occupied by each of the two types of ions. The coordination number of either type of ions is 6. The sulphur ions occupy the corners of the greater cube and the face centres, while the lead ions occupy the interspaces. The pattern may be reversed without changing the nature of the structure. Mostly cubic, sometimes with octahedral faces, less often octahedral. The most common forms are: {100}, {111}, and less often {110}. Twins on (111) as the composition plane. Galena crystals are found only in druse cavities. Usually granular massive or in the form of irregular segregations.

Colour: Lead grey *streak* greyish black. *Lustre,* metallic. *Hardness,* 2 to 3. Brittle, Cleavage, very perfect, cubic. Bismuth-bearing varieties display parting parallel to {111}, which disappears on heating and the usual cubic cleavage appears. Specific gravity, 7.4 to 7.6. Other properties, poor conductor of electricity but has good rectifying properties.

Diagnostic features: Easily recognisable by colour, lustre, characteristic cubic cleavage, low hardness and low specific gravity. In crypto-crystalline masses is distinguished from similar antimonides and arsenides by specific gravity and behaviour before the blowpipe. Easily fusible in the blowpipe flame. Roasted with soda, gives a lead globule.

Readily soluble in nitric acid, giving sulphur and white $PbSO_4$ precipitate due to partial oxidation during solution. Genesis and occurrence, almost exclusively confined to *hydrothermal* deposits. Often found in huge masses. Typically is associated with sphalerite (ZnS), the amount of which is usually greater than that of galenite. Hydrothermal lead-zinc deposits occur either as regular veins or as irregular metasomatic bodies in limestones or, lastly, as disseminated ores. Other galena associates are pyrite, chalcopyrite, grey ores, the sulphosalts of silver, zinc, copper, and arsenopyrite. Apart from quartz and calcite, the non-metallic minerals in these ores are barite ($BaSO_4$), fluorite (CaF_2) and various carbonates.

In the course of oxidation in the process of weathering, galena is covered with an anglesite ($PbSO_4$) crust which alters to cerussite ($PbSO_3$) at the surface. These poorly soluble compounds form a sort of solid coating around undecomposed galena protecting it from the oxidising agents. Therefore, it is not surprising that galena nodules protected by such a coating are found in the zone of clayey alluvium and even in placers. Unlike sphalerite, galena in the zone of oxidation gives rise not only to anglesite and cerussite, but also to other poorly soluble oxygen compounds such as phosphates, arsenates, vanadates and mo-lybdates. Owing to this, the oxidised zones of lead-zinc deposits are, as a rule, enriched in lead.

The most important of the numerous galena ore deposits of the USSR are the following: the *Sadon* vein deposit, Northern Caucasus; the *Turlan* (Achisay), unusually coarsely crystalline massive galena occurs in the limestones of the Kara-Tau range, northeast of the town of Turkestan; the *Altai* polymetallic ore deposits composed of very fine-grained pyrite, sphalerite, chalcopyrite, galena, and grey ores; numerous deposits in Kara-Mazar mountains, and elsewhere in *Central Asia* the *Nerchinsk* deposits in the Trans-Baikal Region, etc.

In *Missouri* (U.S.A.) large deposits occur in the form of impregnations and pockets in limestones and shales over a vast territory; other deposits are worked in *Leadville*, Colorado, and elsewhere.

Uses: Galena is the most important ore of lead. Practically all of the world's production of lead is obtained from this ore. Apart from being used for the smelting of lead, the applications of which are well known, a small amount of galena ores are processed into litharge (PbO) to obtain pigments (white, red, and yellow lead) and glazes. Considerable quantities of silver (and occasionally some bismuth) are obtained as by-product of lead smelting from argentiferous minerals associated with galena.

Sphalerite Group

Dimorphous *AX-type* minerals which crystallise in the cubic and hexagonal systems (crystal structures of the sphalerite and wurtzite type with tetrahedral ionic coordination) are included under this heading. In this section we shall also describe mercury sulphide (cinnabar) which, though close to the sphalerite group in some properties, materially differs from it in crystal structure.

Sphalerite, Zns. The name comes from the Greek "sphaleros" meaning deceptive, probably because it bears no outward resemblance to common metal sulphides. Synonym: zinc blende. Varieties: cleio-phane which is light-coloured or colourless (almost without impurities); marmatite which is black and ferruginous; and pribramite which is rich in cadmium (up to 5 per cent).

Chemical Composition: 67.1 per cent, S 32.9 per cent. The most common impurity is Fe (up to 20 per cent); upon microscopic examination, this variety shows minute inclusions of pyrrhotite (FeS) which is a product of the disintegration of solid solution. Sometimes microinclusions of chalco-pyrite ($CuFeS_2$) and, occasionally, of stannite (Cu_2FeSnS_4) are also found, which explains the presence of copper and tin admixtures in sphalerite. Often other isomorphous admixtures are present in sphalerite. These are Cd (usually up to tenths of one per cent), In (up to hundredths of one per cent), Ca, Mn, Hg, etc.

Crystal structure is characterised by treble-layer (cubic) closest packing of sulphur anions. The zinc cations occupy half the tetrahedral spaces between the anions. This type of structure is similar to that of diamond, the difference being that the centres of the smaller cubes are occupied by atoms (ions), which are different from those at the corners and the face centres of the larger cube. In the unit cell which is shown diagrammatically, there are four sulphur ions at centres of half the number of smaller cubes.

It should be noted that all the tetrahedra are uniformly oriented, which results in tetrahedral rather than cubic symmetry. In contrast to diamond, cleavage in sphalerite crystals is parallel not to the octahedral planes but to the rhombododecahedral planes {110}, since these planar nets contain Zn and S ions simultaneously and in equal numbers.

Habit: Well-developed sphalerite crystals often occur in druse cavities. The tetrahedral habit is the most common, the positive and negative forms often differing in lustre and etch figures. Sometimes {110} faces prevail and the crystals acquire a dodecahedral habit. Commonly twinned along {111}. *Aggregates:* Compact masses show pronounced granular structure, easily recognisable by the very distinct cleavage of the individual grains. Botryoidal forms are less common.

Colour: Usually brownish to brown; often black (marmatite), less often yellow, red or greenish. Perfectly colourless varieties (cleiophane) are also known, streak, white or light yellow to brown. Iron-rich varieties have a brown streak. Lustre, adamantine. Refractive index in Na-light $n = 2.37$. *Hardness,* 3 to 4. Rather brittle. *Cleavage,* {110} excellent. Specific gravity, 3.9 to 4, non-conductor of electricity, polar pyroelectric. Certain varieties phosphoresce on rubbing or splitting.

Diagnostic features: Characteristic isometric crystal grains with rhombododecahedral cleavage, i.e., in the six directions corresponding to the structural planar nets composed of Zn and S atoms. This distinguishes the ferruginous varieties of sphalerite from wolframite (Fe, Mn) WO_4 and enargite, Cu_3AsS_4, which are similar to the former in colour, hardness, lustre and other features, but have a prismatic habit of grains and unidirectional cleavage.

Decrepitates in the blowpipe flame but practically does not fuse. On charcoal in the reducing flame gives a white coating of zinc oxide. Dissolves in concentrated nitric acid with separation of sulphur.

Genesis and Occurrence. Most of sphalerite deposits, as those of galena, with which sphalerite is practically always associated, are of the *hydrothermal* type. In certain sulphide deposits, it is sometimes associated with chalcopyrite.

Under exogenous conditions sphalerite is formed extremely rarely. It has been found in some coal seams. On oxidation sphalerite decomposes rather rapidly with the evolution of zinc sulphite which is readily soluble in water; that is why the oxidation zones are, as a rule, heavily depleted of zinc (cf. galena). When the enclosing rocks of a deposit are limestones, zinc carbonate (smithsonite) will accumulate in them.

In lead-zinc deposits, examples of which have already been cited (under "Galena"), sphalerite usually occurs in much greater quantities than galena. In some deposits, there are found druses of well-developed crystals of sphalerite together with calcite, quartz, and other minerals, for example, in the deposits of the *Nagolny ridge,* the Ukraine.

Sphalerite may also occur as metacolloidal concentrically-banded botryoidal formations in cavities in limestones, in association with galena, pyrite, marcasite, chalcopyrite, calcite, or dolomite.

In close association with chalcopyrite (almost without galena) it occurs in a number of pyrite deposits in the Urals, such as the Xarpu-*shino* and the *IIIrd International* deposits, and elsewhere. Outside the USSR, the most interesting deposits, from the mineralogical standpoint, are *Pribram* (Czechoslovakia), *Binnenihal* (Switzerland) where excellent crystals occur in cavities in dolomite, and the *Santander* deposits (Spain) where remarkable transparent sphalerite crystals have been found.

Uses: Sphalerite is the principal ore of zinc. Apart from zinc, sphalerite ores yield, as by-products, valuable rare metals: Cd, In, and Ca. In the course of roasting and smelting of polymetallic ores, ZnS, oxidised to ZnO, escapes in large amounts with the flue gases. For this reason, the ores are usually pretreated to separate the lead and the zinc concentrates. The latter, after preliminary roasting in special furnaces in order to oxidise the zinc, is smelted in a reducing atmosphere in sealed retorts so as to sublimate the zinc.

The metallic zinc which is obtained by sublimation is not pure and is used to galvanise iron. Crude zinc is refined electrolytically and goes into brass, bronze, and other alloys. Small quantities of sphalerite are used directly for making zinc white, fluorescent screens, etc.

Cadmium, the bulk of which is obtained from sphalerite ores as a by-product, is used in electroplating, as an anti-corrosion coating for steel and iron; in the preparation of fusible alloys which are stronger and more resistant to high temperatures and wear than babbit alloys which require a great admixture of costly tin; in storage batteries; in the manufacture of automatic fire-fighting equipment.

Gallium is a metal, which has many properties in common with aluminium. It melts at 29° and forms an alloy with aluminium, which is liquid at room temperature. The boiling point of gallium, unlike that of mercury, is very high (1700 to 2300°) making it in some applications a good substitute for mercury in thermometers and other precision instruments. Gallium lamps emit light closely similar to daylight.

Indium is used as an anti-corrosive coating for metals and in the manufacture of reflectors for spotlights and automobile headlights. Organic compounds of indium are used in medicinal preparations for the treatment of sleeping sickness.

WUBTZITE, ZnS. ***Variety:*** erythrozincite, i.e., manganiferous wurtzite, (Zn, Mn)S.

Chemical composition: identical with that of sphalerite. Usually contains more cadmium.

Crystal structure is characterised by hexagonal (double-layer) closest packing of sulphur anions. As in the sphalerite structure, the zinc cations occupy half the vacant tetrahedral spaces. This similarity in crystal structure explains the similarity in a number of physical properties (specific gravity, hardness, colour, etc.). However, wurtzite differs essentially from sphalerite by optical anisotropy and crystal form.

Habit: pyramidal, short-columnar or tabular.

Colour: similar to that of sphalerite, but varies with Fe content from white to brown and black. Streak accordingly varies from white to brown. *Lustre,* adamantine. *Hardness,* 3.5 to 4. Brittle. *Cleavage,* {1120} perfect and {0001} indistinct. Specific gravity, 4.0 to 4.1.

Diagnostic Features: Massive wurtzite is outwardly indistinguishable from sphalerite. Its optical anisotropy is revealed only by microscopic examination. Behaviour before the blowpipe and reactions with acids are the same as those of sphalerite.

Genesis and Occurrence: Wurtzite is a comparatively rare mineral. It occurs together with sphalerite in certain hydrothermal deposits of low-temperature origin.

In the USSR, it occurs together with sphalerite in some of the Urals deposits: *Blyava, Yaman-Kasy;* and in Kirghizia: at *Ak-Tyuz, Karavash,* and elsewhere.

Most interesting mineralogically are the concentrically-banded metacolloid varieties in Pribram, Czechoslovakia, the well-formed pyramidal crystals of the *Butte* district of Montana (U.S.A.) and at *Oruro* and *Potosi* (Bolivia). The mineral is of no economic value.

Greenockite, CdS Synonym: cadmium glance. Rare. Cadmium content 77 per cent. Sometimes contains indium. System, hexagonal; symmetry, dihexagonal pyramidal. Space group identical with that of wurtzite, a_0 = 4.142; c_0 = 6.724.

Recently there has been found a cubic modification of this compound which has been called *hauleite.* Rare crystals are small and barrel-shaped or acute pyramidal in form. Usually occurs as powdery or earthy selvages.

Colour: yellow, orange-yellow or dark orange. Streak, orange-yellow to brick red. Lustre, adamantine. Hardness, 3 to 3.5. Brittle. Cleavage, {1120} perfect. Specific gravity, 4.9 to 5.0. Diagnostic features. Differs from similar orpiment, As_2S_3, realgar AsS, and wulfenite $PbMoO_4$, by behaviour before the blowpipe (on strong ignition with soda gives a red-brownish film of CdO). When dissolves in acids gives of a strong odour of H_2S. The reaction for cadmium is performed in an alcohol solution of diphenylcarbazide on filter paper.

Genesis and Occurrence: Is found in association with cadmium-bearing sphalerite or wurtzite. No large accumulations of greenockite have been observed. Has occurred in the oxidised zones of the following deposits: *Kyzyl-Espe* (Central Kazakhstan), Gaynakh-Kan, Kan-Sai and Obisor-bukh (Central Asia). Has been also found in cavities as tiny crystals on chalcopyrite and galena in a pyrite bed of the *Sibai* deposit, Southern Urals, where it is obviously in exogenous mineral.

Other localities *are Pribram,* Czechoslovakia; *Friedensville,* Pennsylvania (U.S.A.), etc.

Cinnabar, HgS: The name presumably comes from India, where it is applied to a red resin and to "dragon blood". A cubic modification of HgS is known as metacinnabarite.

Chemical Composition: Hg 86,2 per cent, S 13.8 per cent. Foreign elements are usually mechanical admixtures.

Crystal Structure, Hexagonal: On the whole, it may be regarded as a distorted structure of NaCl with a coordination number of 6 (both for Hg and for S). A peculiar feature of the crystal structure are continuous spiral chains of S —Hg—S with covalent bonding between the ions, the chains being parallel to the *c* axis (along the right- or left-hand helical axis). This explains the pronounced ease of rotation of the plane of polarisation. The weaker bonding between the chains accounts for the distinct prismatic cleavage parallel to {1010}.

Habit: Cinnabar occurs in small thick tabular (0001) or rhombohedral crystals with faces {1011}, {2025}, etc., and sometimes with trapezohedral faces. Characteristic twinning along (0001). *Aggregates:* Most often observed as impregnated irregular grains, sometimes massive, and also as powdery selvages and coatings. The so-called "hepatic" ore is a cryptocrystalline mass with abundant earthy and organic impurities.

Colour: red, sometimes with a lead-grey tarnish. Streak, red. Lustre, strong, sub-metallic. Translucent. Hardness, 2 to 2.5. Brittle. Cleavage, {1010} rather perfect. Specific gravity, 8.09. Unlike metacinnabarite, cinnabar does not conduct electricity.

Diagnostic Features: Easily identified by red colour, low hardness, high specific gravity, and behaviour before the blowpipe. Volatilises completely in the blowpipe flame on charcoal. Sublimation begins at 200°C. When heated in the closed tube, gives a black sublimate of HgS, metallic mercury, and a sulphur coating. When heated in the open tube, i.e., in the presence of oxygen, tiny globules of metallic mercury are formed on the cool walls of the tube according to the equation: $HgS+O_2=Hg+SO_2$. This is essentially the method employed in industry to obtain mercury.

Cinnabar is soluble in aqua regia and is decomposed by chlorine. It also dissolves in caustic alkali solutions of sulphides, but is insoluble in nitric and sulphuric acids. Genesis and occurrence cinnabar deposits, without exception, belong to the low-temperature *hydrothermal* type. Instances are known of cinnabar deposited from hot alkaline springs at the daylight surface (Steamboat Springs in Nevada, and Sulphur Bank Springs in California, USA). The ore minerals associated with cinnabar are: antimonite (Sb_2S_3), pyrite, less often arsenopyrite (FeAsS), realgar (AsS), occasionally sphalerite, chalcopyrite, etc. The usual non-metallic associates are quartz, calcite, often fluorite and barite, sometimes gypsum, etc.

In the oxidation zones of mercury deposits, black films of metacinnabarite, native mercury, and occasionally mercury chlorides occur sometimes as secondary minerals. In general, cinnabar is rather stable in oxidising conditions, in contrast to many other sulphides. This is why-it is often found in placers where due to its high specific gravity, it shows in the heavy concentrates.

The Soviet Union's largest deposit is at *Nikitovka* (3 km from the Nikitovka Station, Northern-Donets railway) where cinnabar occurs both as an impregnation and in veinlets, in association with quartz, antimonite, arsenopyrite, and occasionally with pyrite, mostly in sandstones. A number of small deposits are situated in the Caucasus. Of these the *Khideshlep* deposit is interesting mineralogically, since there cinnabar is associated with bright-red realgar (As_2S_3).

Larger deposits are found at *Khaidarkan* and *Chauvai* in Central Asia, chiefly along the northern foothills of the Altai and Turkestan ranges. There cinnabar in the form of small grains in brecciated ores is associated with quartz, antimonite, fluorite, calcite, barite, and other minerals. World famous are the deposits at *Almaden* (Spain), and *Idria* (Italy).

Cinnabar is practically the only source of metallic mercury since native mercury is rather rare. The principal use of mercury is for the amalgamation of gold in extracting the latter from primary ores. It is also used for the manufacture of chemicals, fulminate of mercury Hg(CNO) for detonating high explosives, and in scientific instruments.

PYBBHOTITE GROUP

This group includes the compounds of metals of Group VIII of the periodic table (in particular Fe, Ni, and Co) with S, As, and Sb. These compounds have the general formula *AX* (or a very similar formula).

We shall consider here the minerals: pyrrhotite, niccolite, millerite, and pentlandite.

Pyrrhotite, FeS (most often =0.1 to 0.2): Usually its formula is presented as FeS. The name comes from the Greek "pyrrhos" meaning the colour of fire. Synonym: magnetic pyrite.

Chemical Composition: Usually the sulphur content is higher than the one indicated by the formula FeS: instead of 36.4 per cent, it may be as high as 39 to 40 per cent. Sometimes minor impurities of Cu, Ni, Co, and occasionally of Mn and Zn are found (the first three metals are usually present in the form of the chalcopyrite and pentlandite inclusions).

This modification is stable at temperatures below 138°C. The crystal structure is characterised by the hexagonal structure of niccolite. According to X-ray data, the excess sulphur (against the formula FeS) cannot be accounted for by additional large S^{2-} ions fitting somewhere into the empty spaces of the pyrrhotite crystal structure since these spaces are too small. It may be assumed that either S^{2-} ions partly replace

Fe ions or, if the number of S^{2-} ions in the lattice is constant, some of the sites for Fe cations remain vacant. The problem was solved by comparing the calculated specific gravities for either case. The upper curve represents the calculated specific gravities on the assumption that the S ions partly replace the Fe ions; the lower curve represents the other case in which the assumption is that some of the sites for the Fe ions remain vacant; the circles indicate the specific gravities of pyrrhotite varieties actually existing in nature. It is evident from the graphs that the second assumption is correct. Therefore, it should be assumed that in order to neutralise the total negative charge of the S2" anions, some of the Fe ions must have a trivalent positive charge, not a divalent charge. This has been confirmed by very accurate chemical analyses.

Habit: Crystals are generally rare. Usually they are tabular, less often columnar or pyramidal 100 and 101), mostly with the following Jaces: pinacoidal {0001}, prismatic {1010}, dipyramidal {1011}, {2021}, etc.- Twins are rare, with {1011} as the twinning plane. Commonly massive or in the form of irregular disseminated grains.

Colour, dark bronze-yellow with a dark-brown tarnish. Streak, grey-black. *Lustre,* metallic. *Hardness,* 4. Rather brittle. Cleavage, {1010} imperfect. Apart from this, there is a parting parallel to {0001}. Specific gravity, 4.58 to 4.70. Occasionally magnetic. Sulphur-rich varieties ferromagnetic. Good conductor of electricity.

Diagnostic Features: Identifiable by characteristic colour and, often, by magnetic properties.

Fuses in the blowpipe flame into a black magnetic bead. Dissolves with difficulty in nitric and hydrochloric acids, which sharply distinguishes it from troilite (FeS).

Genesis and Occurrence: Pyrrhotite only in rare instances is a high temperature mineral. Its formation, which is the same as that of pyrite (FeS_2), depends not so much on temperature as on the concentration of sulphur in solutions:

> At high S^{2-} concentration, the iron precipitates as a sulphide (FeS_2), while at low concentration a monosulphide (FeS) is formed.

Pyrrhotite is almost exclusively confined to endogenous deposits of diverse genesis. It is quite widespread in basic igneous rocks, chiefly in norites, and sometimes in gabbro-diabases, where pyrrhotite is the principal mineral in sulphide accumulations occurring in close association with pentlandite and chalcopyrite (deposits of copper-nickel sulphide ores).

Sometimes forms large masses in contact-metasomatic deposits, mainly at the boundary with limestones. Chalcopyrite, pyrite, magnetite, blackjack, arsenopyrite, sometimes cassiterite (SnO_2) scheelite ($GaWO_4$), calcite and quartz occur in paragenesis with pyrrhotite. All of them are formed in the later stages of skarn for crystal formation.

In the USSR, the best known contact-metasomatic deposits are the *Bashmakovskoye* and *Bogoslovskoye* in the Turya group of mines (Northern Urals).

In a number of typical hydrothermal deposits pyrrhotite occurs associated with sphalerite, galena, chalcopyrite, cassiterite, arsenopyrite, ferruginous chlorites, carbonates, etc., being the latest to be formed of all these minerals. Well-formed crystals are usually found in druse cavities as a later growth on such earlier minerals as sphalerite, quartz, calcite, etc.

Rare finds of unusual pyrrhotite concretions in association with siderite were recorded from sedimentary rocks at the Kerch iron deposit, and also in phosphorite nodules.

In conditions of weathering in the zone of oxidation, it is the most easily decomposed sulphide. At first, ferrous sulphate is formed and this changes in the presence of oxygen to ferric sulphate. The latter hydrolyses to form insoluble ferrous hydroxides (limonite) and free sulphuric acid which passes into solution.

Uses: Deposits of massive pyrrhotite ores which do not contain other metallic minerals are of limited economic value. As stock material for the production of sulphuric acid they are greatly inferior to pyrite ores. Their sulphur content usually does not exceed 30 to 32 per cent, while in pyrite ores it may be as high as 45 to 50 per cent.

Niccolite, NiAs. Synonym: copper nickel.

Chemical Composition: Ni 43.9 per cent; As 56.1 per cent. Impurities Fe (up to 2.7 per cent), S (up to 5 per cent), sometimes Sb and Go. The commonly observed broad variations in the composition of niccolite are often due to inclusions of foreign minerals which can be seen under the microscope.

System, hexagonal; symmetry, dihexagonal-dipyramidal. Crystal structure, hexagonal with closest packing of arsenic atoms and with nickel atoms occupying all the octahedral spaces. Each arsenic atom is surrounded by six nickel atoms forming a trigonal prism, whereas each nickel atom is surrounded by six arsenic atoms forming an octahedron. In addition, each arsenic atom is close to two nickel atoms (vertically) which are also its closest neighbours. This explains the partly ionic and partly metallic bonding of atoms in this type of crystalline- structures. It also explains not only the higher reflectivity and electrical conductivity, but also the certain variability of the composition of compounds which form crystal structures of this type.

Habit: Crystals are extremely rare and are poorly developed, with predominant {1010} faces. Usually massive, sometimes in botryoidal and other forms.

Colour, pale copper-red. Streak, dark brownish black. Lustre, metallic. *Hardness*, 5. Brittle. *Cleavage*, {1010} imperfect. Specific gravity, 7.6 to 7.8. Good conductor of electricity. Diagnostic features. Readily identifiable by pale copper-red colour and metallic lustre.

Fuses on charcoal before the blowpipe into a brilliant brittle globule and gives off a strong garlic odour. If intensely heated in the closed tube, forms an arsenic mirror on the cool walls. Solution in nitric acid is apple green which turns blue on addition of ammonia. With dimethylglyoxime yields a thick pink precipitate.

Genesis and Occurrence: Occurs mostly in *hydrothermal* vein deposits, sometimes in large amounts, as impregnations or block masses. Common paragenetic associates: diarsenides of nickel, such as chloan-thite, rammelsbergite, sometimes native bismuth and silver, etc.

The weathering of niccolite gives rise to bright-green annabergite, $Ni_3[AsO_4]_2.8H_2O$.

In the USSR, most interesting finds of niccolite were made at the *Berikul* primary gold deposit, Western Siberia. Together with other nickel arsenides (rammelsbergite, gersdorffite, etc.), it was found as continuous irregularly shaped pocket-like masses in carbonate veinlets.

Other notable occurrences are the hydrothermal vein deposits of the *Erzgebirge mountains* (Saxony) of the so-called cobalt-nickel-silver formation with native bismuth (Schneeberg type), and the *Cobalt deposit* in Ontario, Canada, where it occurs in association with sulphides and arsenides of nickel and cobalt, as well as with native silver and other minerals.

Uses: Niccolite ores when found in quantities are of great economic importance.

Millerite, NiS. Synonym: capillary pyrite. Chemical composition. Ni 64.7 per cent, S 35.3 per cent. Impurities: Fe (up to 1-2 per cent), Co (up to 0.5 per cent), Cu (up to 1 per cent). System trigonal; Crystal structure differs from the structures of pyrrhotite and the artificial modification of NiS (niccolite type with coordination number 6). The structure of this modification with coordination number 5 (intermediate between the high-temperature NiS modification and pentlandite) is very complex and will not be described here.

Habit: Crystals usually acicular with coarse lengthwise striation. Aggregates. Often in radiating hair-like groups, hence the name of capillary pyrites.

Colour, brass yellow, sometimes with an iridescent tarnish. Streak, greenish black. Lustre, strong metallic. *Hardness*, 3 to 4. Brittle. Capillary crystals somewhat elastic. *Cleavage*, {1011} and {0112} perfect. Specific gravity, 5.2 to 5.6. Good conductor of electricity.

Diagnostic Features: Identifiable by very characteristic frequent needle-like forms and radially fibrous brass-yellow aggregates. When found in irregularly shaped grains or masses, requires for identification chemical tests for nickel and sulphur.

Fuses in the blowpipe flame on charcoal to give a brilliant boiling bead. In the reducing flame it finally gives a compact, slightly magnetic nickel globule.

Dissolves in nitric acid and aqua regia colouring the solution green (due to the presence of nickel) with the separation of sulphur. With dimethyl glyoxime gives the characteristic nickel reaction.

Genesis and Occurrence: A relatively rare mineral, it is nearly always a typical product of hydrothermal processes.

Sometimes occurs in deposits of copper-nickel sulphide ores as a later hydrothermal mineral formed at the expense of pentlandite.

In typical hydrothermal vein deposits is found in association with other nickel and cobalt minerals, mostly sulphides and arsenides. In these cases it is observed in radial groups in paragenesis with linneite, gersdorffite, galena, fluorite, calcite, quartz, etc.

In the USSR, mineralogically important finds of millerite

have been recorded from the area of the *Beryozovskoye* gold fields as radiated and sheaf-like aggregates in thin carbonate veinlets in listvenites (hydrothermally altered ultrabasic rocks).

In other countries the best known are the deposits in the *Erzgebirge mountains* (Saxony), namely *Freiberg, Schneeberg,* etc., where millerite associates with other sulphides of nickel and cobalt, and also with galena, calcite, fluorite, etc. Fine-radial aggregates of millerite have been found in the coal mines of Kladno, Czechoslovakia.

Uses: As a nickel-rich mineral, millerite is important economically even when found as a sparse dissemination in rocks or ores, especially if associated with other nickel or cobalt minerals.

Pentlandite, $(Fe, Ni)_9S_8$: Synonym iron-nickel pyrite.

Chemical Composition Variable: Fe-Ni ratio is usually 1:1. Cobalt is always present in some quantity (0.4 to 2.5 per cent or more) as an isomorphous admixture to nickel.

System, cubic; *symmetry,* hexoctahedral. Has never been found in well-formed crystals. Occurs in pyrrhotite ores of magmatic deposits of the Sudbury type as irregular grains and inclusions. Crystal structure. The sulphur anions are arranged in cubic closest packing. The iron and nickel cations in the main occupy half the tetrahedral interstices (the number of which equals that of the sulphur anions), and some of them (according to the chemical formula) occupy one eighth of the octahedral interstices.

Colour: bronze-yellow, somewhat lighter than that of pyrrhotite. Streak, greenish black. Lustre, metallic. *Hardness,* 3 to 4. Brittle. Cleavage, octahedral {111}, perfect. Specific gravity, 4.5 to 5. Not magnetic. Good conductor of electricity.

Diagnostic Features: Identifiable with difficulty without the aid of the microscope, since it usually occurs as minutest segregations in a pyrrhotite groundmass. Only large grains can be distinguished by their somewhat lighter colour as compared with pyrrhotite and clearly defined cleavage.

Before the blowpipe it fuses into a black magnetic globule. Dissolves in HNO_3 imparting a green colour to the solution. A dark-brown precipitate of ferrous hydroxide is formed on addition of ammonia. Gives the characteristic nickel reaction with dimethyl glyoxime.

Genesis and Occurrence: Almost always occurs in paragenesis with pyrrhotite and chalcopyrite, but only in those sulphide ores which are associated genetically with basic and ultrabasic igneous rocks (gabbro-norites, peridotites, etc.). The paragenesis of the three minerals in such rocks is so characteristic that it is enough to establish in them the presence of the more easily identifiable pyrrhotite and chalcopyrite to be sure that microscopic examination will reveal the presence of pentlandite, which is of great economic importance. Magnetite and platinum minerals such as sperrylite—$PtAs_2$, palladium platinum, sulphide— PtS and braggite—(Pt, Pd, Ni) S are also found in small quantities.

In the zone of oxidation, nickel sulphides alter to nickel sulphate which is readily soluble in water, and is often observed in cavities and on the walls of mine workings as pale-green stalactites and crystalline encrustations of the following composition: $NiSO_4.7H_2O$ (morenosite) or $NiSO_4.6H_2O$ (roetgersite).

The largest deposit of such ore is at *Sudbury* in Ontario, Canada. The ore bodies in the form of large beds and veins lie in the lower parts of magmatic masses composed of basic rocks (norites, gabbro, etc.), and also in the rocks underlying the metamorphic complex. The ore contains: Ni 1 to 5 per cent, Cu and platinum metals 2 to 3 per cent.

Uses: Pentlandite ores are the chief source of nickel. About 90 per cent of the world's nickel production comes from copper-nickel sulphide ores (mostly from Sudbury). In addition to nickel, extracted from these ores are cobalt, copper, platinum metals, and in small quantities, selenium and tellurium.

Nickel is used in the manufacture of special tools, utensils, in the production of many important alloys (white copper, nickel steel, and copper-zinc alloy for rheostats, coinage, etc.) in nickel-plating compounds, etc.

Chalcopybite Group

Under this group we shall describe the complex sulphides of Cu, Fe, and Sn—chalcopyrite and stannite—which crystallise in the tetragonal system. Although very similar to sphalerite in crystal structure, they differ sharply from it in physical properties. These sulphides sometimes occur in sphalerite as minute inclusions resulting from the disintegration of a limited solid solution, which may be explained by the close similarity of the crystal structures, especially at high temperatures. In turn, sphalerite may be present in stannite as a product of disintegration of a solid solution.

Bornite and cubanite also belong to this group, the former being a complex copper-iron sulphide very similar to chalcopyrite with which it can form a wide variety of isomorphous mixtures, although these break up upon cooling.

Chalcopyrite, $CuFeS_2$: "Chalcos" is the Greek for copper, "pyros" fire. Synonym: copper pyrite.

Chemical Composition: Cu 34.57 per cent, Fe 30.54 per cent, S 34.9 per cent. The data of chemical analyses usually closely agree with these figures. Negligible admixtures of Ag, Au, etc., are sometimes present.

System, tetragonal; *symmetry,* tetragonal-scalenohedral. Crystal structure is a relatively simple, nearly cubic, tetragonal lattice. The unit cell of chalcopyrite may be described as that of the sphalerite type doubled vertically. As in sphalerite, each sulphur ion is surrounded by four metallic ions which are located at the corners of a tetrahedron, i.e., ions of copper and iron are spaced in each sheet after a definite pattern. In the first and fifth cationic sheets, i.e., on the upper and lower faces of the tetragonal prism, Fe ions occupy the corners of the square, while Cu ions occupy the centre. In the third sheet (in the middle of the prism) the arrangement is reversed: Cu ions occupy the corners of the square, while Fe ions are at the centre. In the second and fourth sheets, two Cu ions intersect with two Fe ions, the Fe cations being beneath the Cu cations of the second sheet and vice versa. All the tetrahedral groupings have the same orientation which accounts for the crystals being hemi-hedral. In distinction from sphalerite, chalcopyrite is opaque, has a clearly metallic lustre, and does not possess perfect cleavage.

Habit: Crystals' are rare and occur only in druse cavities. Mostly octahedral with combinations {112} and {112}, or tetrahedral; less often scalenohedral, etc. The faces of the basic tetrahedron are either dull or striated, while the negative faces are smooth. Often twinned, with twinning plane (112), less often (102).

Aggregates: Usually massive or in irregular disseminated grains. Also occurs in collo-form kidney-shaped and grape-like forms.

Colour, brass yellow often with dark-yellow or iridescent tarnish. *Streak,* black with a greenish tint. Opaque. *Lustre,* strong metallic. *Hardness,* 3 to 4. Rather brittle. *Cleavage,* {101} imperfect. Specific gravity, 4.1 to 4.3.

Diagnostic features: Readily identifiable by colour; hardness very different from that of pyrite which on fracture often shows an iridescent tarnish, somewhat similar to the colour of chalcopyrite. Irregularly-shaped grains of millerite look like chalcopyrite but have a stronger lustre and a higher nickel content.

In the blowpipe flame decrepitates and fuses to a magnetic globule. Roasted with soda on charcoal, gives a copper globule. In the closed tube, it gives a sulphur sublimate. In nitric acid gradually decomposes with separation of sulphur.

Genesis and occurrence: In nature, chalcopyrite forms under a variety of conditions. As an accessory of pyrrhotite, it often occurs in deposits of copper-nickel sulphide ores of *magmatogene* origin in basic igneous rocks in association with pentlandite, magnetite, sometimes cubanite, etc.

It most commonly occurs in typical *hydrothermal* vein and metasomatic (including contact-metasomatic) deposits. Is often associated with pyrite, pyrrhotite, sphalerite, galena, grey ores and many other minerals. The non-metallic minerals in these deposits are quartz, calcite, barite, silicates of diverse composition, etc.

Chalcopyrite is very rarely the product of *exogenous* processes in sedimentary rocks forming in the conditions of hydrogen-sulphide fermentation in the course of decomposi tion of organic remains and an influx of cupriferous solutions. Along with chalcocite and marcasite it sometimes replaces wood fibres and other organic remains.

As a result of chemical weathering chalcopyrite yields copper and iron sulphates. Interacting with CO_2, or with carbonates in the presence of oxygen and water, soluble copper sulphate forms malachite and azur-ite; reacting with SiO_2 hydrosols, it gives chrysocolla; whereas reactions with different acids formed in the zone of weathering give rise to various salts: arsenates, phosphates, vanadates and, sometimes, chlorides. However, in a very dry climate various copper sulphates survive in the oxidised zone despite their ready solubility in percolating surface waters.

Pseudomorphs after chalcopyrite, i.e., its replacement by secondary copper sulphides, such as bornite, chalcocite and covellite, are common in the zones of secondary sulphide enrichment of copper deposits.

Chalcopyrite, as an associate, occurs in varying amounts in practically all the hydrothermal deposits of most diverse sulphide ores. It forms a substantial part of the ores of many deposits and is therefore of economic value. The USSR has deposits of all genetic types in which chalcopyrite is the principal copper mineral.

Other important localities for chalcopyrite are *Bingham,*

Utah (U.S.A.); *Chuquicamata,* Chile; the cupriferous sandstones of *Katanga,* Congo, and Northern Rhodesia, south of Katanga.

Uses: Chalcopyrite ores are a major source of copper. The average copper content usually ranges between 2 to 2.5 per cent. The copper smelted from the ores is used either pure or in the form of alloys (brass, bronze, tombac, etc.). The chief use of copper is in electrical engineering. Other major uses are in the machine-building and shipbuilding industries, in the manufacture of chemical equipment, plumbing, etc.

Stannite, Cu_2FeSnS_4. Synonym: tin pyrite.

Chemical composition: Cu 29.5 per cent, Fe 13.1 per cent, Sn 27.5 per cent, S 29.9 per cent. Content according to analyses: Sn 22.0 to 27.7 per cent, Cu 22.9 to 31.5 per cent, Fe 4.7 to 23.3 per cent. Impurities: Zn 0.75 to 10.1 per cent, Sb up to 3 per cent, Cd up to 1.5 per cent, Pb up to 2 per cent, Ag up to 1 per cent.

System, tetragonal; *symmetry,* tetragonal-scalenohedral. Rare small crystals have cubic or tetrahedral habit. Externally very similar to chalcopyrite crystals. Usually occurs as irregular grains or massive. Crystal structure, perfectly identical with that of chalcopyrite. The cations are arranged as follows: in the first and fifth sheets, Sn ions are at the corners of the square, while Fe ions are at the centre; in the middle sheet, the reverse is observed—Fe ions are at the corners, while Sn ions are at the centre; the even sheets (second and fourth) are composed of Cu ions. Thus, the crystal structures of stannite and chalcopyrite are very similar in ionic arrangement to that of sphalerite.

Colour, steel grey with a characteristic olive-green tint on fresh fracture. When numerous microscopic chalcopyrite inclusions are present, stannite becomes definitely yellowish. Streak, black. Opaque. *Lustre,* metallic upon fresh fracture but quickly turns dull. *Hardness,* 3 to 4. Brittle. *Cleavage,* {110} and {001} imperfect, rarely observed. Specific gravity, 4.3 to 4.5.

Diagnostic Features: Has a characteristic colour with an olive-green tinge and is readily distinguished by the naked eye from grey ores which are similar to stannite in a number of properties (hardness, brittleness, etc.).

Fuses before the blowpipe on charcoal, turning white at the surface, and giving a white coating of SnO_2 around the assay. The presence of copper, iron, and sulphur is determined by chemical reactions. Decomposes in nitric acid with separation of sulphur and stannous oxide, imparting a deep blue colour to the solution.

Genesis and Occurrence: Stannite is a comparatively rare mineral and occurs in *hydrothermal* tin ore deposits. In tungsten-tin deposits, it is observed in association with cassiterite (SnO_2) chalcopyrite, arsenopyrite, wolframite, and other minerals. Stannite is much more common in stanniferous sphalerite-galena and sphalerite-pyrrhotite ores. In these it is closely associated paragenetically with sphalerite, chalcopyrite, and sometimes pyrrhotite, galena, etc. Instances are known when it is replaced by cassiterite (SnO_2) and vice versa.

In the oxidation zone, it readily decomposes to give finally limonite and cassiterite. In this process, however, most of the tin apparently passes into colloidal solution which finally coagulates to form earthy, spongy or colloform cassiterite concretions. In the USSR, stannite occurs in numerous stanniferous ore deposits, most often in association with chalcopyrite in grains of up to 1 cm in diameter in paragenesis with cassiterite and sphalerite.

Outside the USSR, it has been found at *Zinnwald* (Erzgebirge mountains, Czechoslovakia) and in considerable quantities in Southern China, in *Tsihan* (Tasmania), and especially in *Bolivia* (Ahota, Potosi, etc.).

Uses: Stannite very rarely is found in considerable quantities and, therefore, it is of lesser importance to industry than cassiterite.

Bornite, Cu_5FeS_4. Synonym: variegated copper ore. In nature, it forms limited solid solutions with chalcopyrite, which disintegrate upon cooling. The process of disintegration has been studied experimentally.

Chemical Composition: Chemical composition is variable. Theoretically, according to the chemical formula Cu_5FeS_4 the composition should be: Cu 63.3 per cent, Fe 11.2 per cent, S 25.5 per cent. However, the composition of bornite varies widely because the mineral may contain chalcopyrite and chalcocite in the form of solid solutions. Another frequently observed chemical impurity is Ag.

System, cubic; symmetry, hexoctahedral. Crystals are extremely rare. Usually massive or in the form of impregna tions. Crystal structure. Complicated cubic. According to X-ray data, two kinds of copper ions occupy different positions and its chemical formula is probably: $2Cu_2SCuFeS_4 = (Cu_4CuFeS_4)$, i.e., four of the copper ions are univalent, while the fifth copper ion and the iron ion are divalent.

Colour, dark copper-red on fresh fracture; usually shows a bright variegated (mostly blue) tarnish. Streak, greyish black. Opaque. *Lustre,* submetallic. *Hardness,* 3. Rather brittle. *Cleavage,* practically none. Specific gravity, 4.9 to 5.0. Other properties. Conductor of electricity.

Diagnostic Features: Easily recognised by colour and variegated, predominantly blue, tarnish, and low hardness. May be mistaken for covellite because of bright blue tarnish (but when scratched with a knife, shows its real colour).

Before the blowpipe bornite fuses into a magnetic bead: roasted with soda on charcoal, gives a copper globule. Being acted upon by HNO_3, it decomposes with separation of floating sulphur.

Genesis and Occurrence: Bornite may be of both endogenous and exogenous origin. Bornite of endogenous origin occurs in certain hydrothermal deposits. It sometimes contains microscopic, usually platy, chalcopyrite inclusions, which are a product of solid solution disintegration. Apart from chalcopyrite, paragenetic associates of bornite include occasionally endogenous chalcocite, galena, sphalerite, pyrite, etc.

Exogenous bornite is widespread in *the zones of secondary sulphide enrichment.* As the earliest secondary sulphide, it is formed metasomatically mostly after chalcopyrite as irregular veinlets, selvages or in massive forms.

Although present in numerous copper sulphide deposits, bornite seldom occurs in quantity. Bornite is less stable than other secondary copper sulphides and is replaced by chalcocite and covellite which are richer in copper. Decomposition of bornite in the oxidation zone gives rise to oxygen compounds: malachite, azurite, less often cuprite, etc.

In the USSR, considerable quantities of endogenous bornite in association with chalcopyrite have been found in the *Uspenskoye* deposit.

Exogenous bornite occurs in large amounts in the zones of secondary sulphide enrichment in almost all copper sulphide deposits, especially under the conditions of weathering in a temperate climate.

Uses: Since bornite contains much more copper than chalcopyrite, even its impregnation ores, if found in sufficient quantities, are definitely worth mining.

Cubanite, $CuFe_2S_3$: System, orthorhombic. Chemical composition: Cu 22 to 24 per cent, Fe 40 to 42 per cent, S 34 to 35 per cent. Colour, bronze yellow similar to that of pyrrhotite. Lustre, metallic. Hardness, 3.5. Cleavage, none. Specific gravity, 4.03 to 4.18. Highly magnetic.

Paragenetically, cubanite is closely associated with chalcopyrite. Is often observed in the latter as platy inclusions visible under the microscope, formed as a result of the disintegration of solid solutions. First discovered in the gold-quartz veins of Morro-Velho, Minas Geraes, Brazil.

Covellite Group

This group includes compounds which have a simple empirical formula, but a complex crystal structure in which each element is represented by two sorts of ions (this applies equally to the cations and the anions). Therefore, the chemical formula appears more complex. We shall consider only one mineral of this group—covellite.

Covellite, CuS, or $Cu_2S\text{-}Cu\text{-}S_2$. Named for the Italian mineralogist Covelli. Synonym: copper indigo.

Chemical Composition: Cu 66.5 per cent, S 33.5 per cent. Chemical analyses reveal impurities of Fe, and, less often, of Se, Ag, and Pb.

System, hexagonal; *symmetry,* dihexagonal dipyramidal. Extremely rare crystals are small, thin and tabular. Crystal structure. Covellite has a peculiar layered hexagonal lattice 105). X-ray studies have shown the structure to be much more complex than the original CuS formula suggests. The sulphur ions proved to be of two kinds: S^{2-} single ions and $[S_2]^{2-}$ coupled ions, i.e., compact pairs of ions which are similar to those found in the crystal structure of pyrite and having the same S —S distance, namely, 2.05 A 115).

The copdivalent copper ion is surrounded by three single S^{2-} ions, forming an equilateral triangle. The triads are connected to each other through the corners to form sheets (shown black in 105) in the crystal structure of covellite, which are oriented perpendicular to the sixfold axis. Between these sheets there are two layers of tetrahedra with univalent ions Cu^{1+} at the centre. The bases of the tetrahedra facing one another are connected by vertically oriented $[S_2]^{2-}$ ion pairs (in the middle, upper, and lower layers, and which are shown in 105 as hollow shortened triangular prisms); the tetrahedral apexes opposite to these are occupied by single S^{2-} ions which are shared by the triangular sheet and the apexes of the next tetrahedral layer.

Thus, the structure of covellite is a combination of the structural elements of both modifications of carbon (diamond and graphite). These structural features agree well with the shape, cleavage and unusual optical properties of the mineral and its ability to yield a part of its sulphur as sublimate on heating, etc.

Aggregates: Covellite usually occurs as thin bright deep blue coatings or crusts or as deep blue-black powdery or sooty masses.

Colour, indigo-blue. Streak, grey to black. Opaque. Translucent and green in extremely thin plates. *Lustre,* metallic. *Hardness,* 1.5 to 2. Brittle. Thin plates somewhat flexible. *Cleavage,* {0001} perfect. Specific gravity, 4.59 to 4.67.

Diagnostic Features: Easily recognisable by deep blue colour, low hardness, and copper sulphide associations. Before the blowpipe covellite fuses readily and burns with a blue flame, evolving SO_2- Unlike chalcocite, it gives a sulphur sublimate in the closed tube. Dissolves in hot HNO_3 with the separation of sulphur.

Genesis and Occurrence: Usually found in small quantities, covellite is a most typical exogenous mineral of the zone of secondary sulphide enrichment in copper deposits. As a rule, it is formed metasomatically after primary and secondary copper sulphides, such as chalcopyrite, bornite, chalcocite, etc. May also be deposited along fissures as colloform or earthy masses.

Covellite of *hydrothermal* origin in paragenesis with pyrite is extremely rare and occurs in small amounts in Butte, Montana (U.S.A.), and elsewhere. As a sublimation product from fumaroles, it is found in the lavas of Vesuvius, where it was first described.

It never forms deposits in its own right, but is found in negligible quantities in virtually all copper sulphide deposits (in the zones of secondary sulphide enrichment). Large masses occur on *Kawau Island* (New Zealand).

Covellite-chalcocite ores are twice as rich in copper as equally disseminated primary chalcopyrite ores.

Orpiment Group

In this group, we shall review a sulphide of trivalent arsenic known as orpiment (auripigment), and also realgar with which it has much in common.

Orpiment, As_2S_3. The name (orpiment or auripigment) comes from the Latin "aurum", gold, and "pigrnentum", paint. It was formerly believed that this mineral contained gold.

Chemical Composition: As 61 per cent S 39 per cent. Usually contains only mechanical impurities: Sb_2S_3, FeS_2 (marcasite), SiO_2, clayey matter, etc. Only Se,Sb,V(upto 0.02 per cent) and Ce (upto $4 \cdot 10^{-5}$ per cent) may be regarded as isomorphous admixtures.

System, monoclinic; symmetry, rhombic prismatic. X-ray data (obtained by V.I. Mikheyev) indicate that certain specimens of orpiment contain considerable amounts of native sulphur as a mechanical impurity. Crystal structure is layered. The layers composed of As_2S_3 are loosely linked by Van der Waals forces, which results in excellent cleavage parallel to (010), low hardness, high optical anisotropy, and other properties. Habit. Crystals usually prismatic 106), often with curved faces. Most common forms are pinacoids {100} and {010}, and prisms {110}, {301}, {210}, etc. Aggregates are often rod-like, or cockscombed, also in botryoidal, reniform, and globular masses with a radial structure.

Colour, lemon yellow to brownish yellow: cryptocrystalline masses, containing admixtures of finely dispersed FeS_2, are dirty yellow with a greenish tint. Streak has the same colour but is somewhat brighter. Translucent; cleavage flakes transparent. *Lustre,* adamantine to sub-metallic, depending on direction. *Hardness,* 1 to 2. Lamellae, flexible but not elastic. *Cleavage,* {010} highly perfect, and {100} imperfect. Specific gravity, 3.4 to 3.5. Non-conductor of electricity. Becomes electrified when separated along cleavage planes.

Diagnostic Features: Easily recognisable by bright lemon-yellow colour, low hardness, excellent cleavage, and strongly adamantine or sub-metallic lustre on fracture. In powdery masses, may be mistaken for pulverulent native sulphur, and also for certain uranophosphates, or uranovanadates, such as autunite and tyuyamunite, from which it can be distinguished by characteristic behaviour before the blowpipe, optical properties, and non-radioactivity.

Easily fuses and volatilises, leaving on charcoal a white coating of As_2O_3 with a sharp arsenical (garlic) odour. Dissolves in nitric acid and aqua regia with separation of floating sulphur, and is readily soluble in KOH, leaving no residue.

Genesis and Occurrence: Occurs in hydrothermal deposits in association with minerals which are formed at comparatively low temperatures, such as realgar, antimonite, marcasite, pyrite, and also with quartz, calcite gypsum, etc. Is deposited from the well-known mineral hot springs in Steamboat, Nevada (U.S.A.) together with realgar, opal, aragonite, etc.

In very small amounts it has been found on the walls of craters, in the cavities in porous lavas as a sublimate together with native sulphur, chlorides, and other minerals.

Exogenous orpiment is very rarely found in negligible quantities, as coatings and earthy accumulations in coal and brown iron ore deposits, being probably the product of the action of hydrogen sulphide of organic origin on arsenious solutions.

In the USSR, orpiment is found in substantial quantities as an arsenic ore in association with realgar. The ores of the *Lukhum* deposit are remarkable for the colouring of hand specimens, which are composed of crystalline masses of bright-yellow auripigment, red realgar, milk-white quartz and calcite. In cavities there often occur druses of well-formed crystals from 0.5 to 5 cm long. Orpiment or auripigment has also been found as large continuous reniform masses, which on fracture show a radial structure.

The auripigment ores of the relatively young *Julfa* deposit (north of the town of Julfa, Nakhichevan Autonomous Soviet Socialist Republic) are quite different in appearance. These occur in tertiary folded marls and argillaceous rocks. Auripigment here was observed in the form of cryptocrystalline (metacolloidal) masses with a waxy lustre on fracture and greenish-yellow colour (due to admixtures of finely-dispersed iron disulphides). During mining operations, abundant evolution of gases, such as CO_2, H_3As, and H_2S, took place. In the vicinity of the deposits, mineral springs containing $Ca(HCO_3)_2$, Na_2SO_4, $CaHAsO_4$, Br, Li, etc., are still flowing.

Other notable auripigment deposits are at *Allachar*, Macedonia, from which large crystals have been recorded, and at *Mercure*, Utah (U.S.A.), and elsewhere.

Uses: When found in quantity is used for obtaining arsenic trioxide, also in dyeing and for other purposes. *Realgar, AsS.*

Chemical Composition: As 70.1 per cent; S 29.9 per cent. Chemical analysis data almost wholly agrees with the theoretical. Isomorphic admixtures of other elements have not been found.

System, monoclinic; *symmetry,* rhombic prismatic. Crystal structure is rather complex, being composed of separate As_4S_4 molecules. The sulphur ions form a square, whereas the arsenic ions form a tetrahedron. The centres of the square and the tetrahedron coincide. Habit. Crystals usually prismatic and shortened or elongated along the vertical axis with thin striations parallel toil on the faces. The most common forms are pinacoids {001}, {010} and orthorhombic prisms {110}, {120}.

Aggregates: Also occurs in massive granular aggregates, sometimes as coatings, crusts or earthy friable masses.

Colour, orange-red, less often dark red, semi-transparent. Streak, light orange. *Lustre,* adamantine on crystal faces, resinous to greasy on fracture. Refractive indices in Li-light: $n_x = 2.61$, $n_y = 2.59$, and $n_z = = 2.46$. *Hardness,* 1.5 to 2. *Cleavage,* {010} and {120} quite perfect. Specific gravity, 3.4 to 3.6. On exposure to light it gradually turns into a light-orange powder. On exposure to electric light the crystals decrepitate and form the same light-orange powder. According to X-ray data, this is accompanied by a breakdown of the crystal structure. Non-conductor of electricity.

Diagnostic features: Characteristic orange-red colour, low hardness, and striation of faces parallel to the crystal elongation axis. Another characteristic feature is its paragenetic association with orpiment, which is easily recognisable by its external features. It differs from similar crocoite ($PbCrO_4$) by lower hardness, different habit, and behaviour before the blowpipe (roasted on charcoal with soda, crocoite gives a lead globule). Cinnabar is distinguished from realgar by its bright-red streak, high specific gravity and different behaviour before the blowpipe.

Fuses readily in the blowpipe flame, and volatilises giving off the characteristic garlic odour. Dissolves in aqua regia with separation of sulphur. Unlike cinnabar, dissolves in hot KOH and gives a lemon-yellow flocculent precipitate on addition of hydrochloric acid.

Genesis and Occurrence: Realgar occurs in nature under conditions absolutely similar to those of orpiment with which it is constantly associated (cf. orpiment). It never occurs at the earth's surface, since on exposure to daylight it decomposes and partly alters to orpiment.

In the USSR, extremely large (1 to 2 cm) realgar crystals 108) and beautiful druses have been recorded from *Lukhum* deposit (Rachinsky ridge on the southern slope of the Caucasus in Western Georgia) in association with orpiment, and occasionally with antimonite, pyrite, marcasite, melnikovite, quartz, ealcite, etc.

As an associate, realgar occurs practically in all orpiment deposits.

Uses: Realgar is a rather rare mineral. In a few cases it forms purely arsenic deposits together with orpiment, one example being the deposit at Lukhum. When thus found, it is a raw material for obtaining As_2O_3 (by roasting). Arsenic sulphide (AsS), both artificial and natural, is used in dyeing, fireworks, in glass making, and in other applications.

Antimonite Group

The sulphides of trivalent antimony and bismuth—antimonite and bismuthinite—which belong to this group fundamentally differ from orpiment in crystal structure and physical properties (opacity, metallic lustre, etc.). This is obviously linked with the higher polarising ability of the Sb^{3+} and Bi^{3+} cations.

Antimonite, Sb_2S_3. From the Latin "antimonium". Synonyms: stibnite, antimony glance.

Chemical Composition: Sb 71.4 per cent, S 28.6 per cent. Impurities: As, Ag, and Au. Compounds of the last two elements are probably present as mechanical inclusions. Native gold has been repeatedly observed in the antimonite groundmass in polished sections under the microscope.

System, orthorhombic; *symmetry,* rhombic-dipyramidal. Crystal structure consists of bands of closely bonded Sb and S ions which are parallel to the *c* axis and which are made up of zigzag chains—Sb—S — —Sb—S —. The bonds between the bands are weaker than those between the ions within the bands. This, naturally, is reflected not only on the shape of the crystals, but also in such properties as cleavage, hardness, brittleness, etc.

Habit: Usually prismatic 109, columnar, acicular, vertically striated. Often, especially in the case of large crystals; curved or even twisted individuals are found. The following combinations are the most characteristic of the many known faces: prism {110}, pinacoid {010}, and pyramids {111}, {113}, {121}, etc.

Aggregates: Occurs as massive granular aggregates, often radial, or, less often, with a felted texture, also as granular impregnations in a quartz groundmass.

Colour and streak, lead grey. Crystals often show a dark bluish tarnish. Opaque. *Lustre,* metallic, and strong on cleavage surfaces. *Hardness,* 2 to 2.5. Brittle. *Cleavage,* {010} perfect, and {110} imperfect. Specific gravity, 4.6. Non-conductor of electricity.

Diagnostic Features: The colour and mechanical properties of aggregates resemble those of many sulphoantimonites, such as boulangerite, jamesonite, and especially bismuthinite. Its distinguishing features are perfect cleavage along the length of columnar individuals, and a cross-wise striations on the shearing surfaces. In the form of fine-granular and cryptocrystalline masses and also when disseminated, it is unmistakably distinguished from all similar minerals by its reaction with KOH. A drop of this reagent placed on the mineral vigorously decomposes it, and the mineral first turns yellow and then orange; when the drop is removed, a red spot will persist.

Very easily fuses in the blowpipe flame on charcoal, giving off SO_2 and leaving a white coating of Sb_2O_3 which vaporises in the reducing flame, imparting a green colouration to the latter. Dissolves in HNO_3 with separation of Sb_2O_5.

Genesis and Occurrence: Antimonite occurs for the most part in *hydrothermal* deposits, being formed at the lowest temperatures, composing, together with quartz, veins and sheet-like bodies. Cinnabar, fluorite, quartz, calcite, kaolinite, barite, etc., are commonly associated with it. It is nearly always an accessory mineral in deposits of cinnabar, realgar, and orpiment, rarely in those of lead, zinc, and other metals. Has been observed in negligible quantities in the sublimates of volcanic eruptions.

In oxidised zones, it decomposes quite readily into various yellow and sometimes dark-brown antimony oxides (valentinite, servantite, kermesite, etc.).

In the USSR, the most important deposits are the *Razdolninskoye*, Krasnoyarsk region, where it occurs as composite quartz-antimonite veins in Pre-Cambrian schists; the *Kadamjay*, 35 km southwest of the town of Ferghana—a sheet-like deposit in "flinty breccia" where antimonite is associated not only with quartz and calcite but also with minor amounts of pyrite, marcasite, fluorite, and barite.

On the Island of Shikoku (Japan) unusually large (up to 0.5 m long and 5 cm thick) 109 antimonite crystals have been found. Great deposits are known in Yunnan and other provinces in China, where antimonite occurs in quartz veins and stock-like bodies in limestones. These deposits are the world's principal sources of antimony ores.

Uses: Antimonite ores are the chief source of antimony—an element which finds most diverse applications. It is used mainly in anti-friction alloys for bearings, such as babbits. Its lead and zinc alloys go to make type, shot, pump and valve parts, etc. Antimony compounds are used in the production of rubber (vulcanisation), textile industry (impregnation of fabrics), glass manufacture, medicine, etc.

Bismuthinite, Bi_2S_3: Synonyms: bismuthine, bismuth glance.

Chemical Composition: Bi 81.3 per cent, S 18.7 per cent. Minor quantities of Pb, Fe, As, Sb, Te, etc., are common admixtures; of these Pb, Sb, and Te may isomorphously replace bismuth.

System, orthorhombic; *symmetry,* rhombic dipyramidal. Crystal structure, similar to that of antimonite. Habit. Like antimonite, occurs in elongated columnar crystals mostly bounded by the prism faces, such as 110}, {120}, {130}, and pinacoids, such as {100}, {010}, {001}. Faces most often show thin vertical striations. Massive, granular, occasionally radial.

Colour, white with a lead-grey tinge. Often with a yellow or iridescent tarnish. Opaque. Streak, grey. *Lustre,* strongly metallic. *Hardness,* 2 to 2.5. *Cleavage,* {010} perfect and {100} and {001} imperfect. Specific gravity, 6.4 to 6.8, sometimes up to 7.1. Non-conductor of electricity.

Diagnostic Features: Bismuthinite differs from antimonite by stronger lustre, higher specific gravity, and reaction with KOH (cf. antimonite).

Aggregates: Resemble numerous sulphoantimonites and sulpho-bismuthinites of complex composition, from which it is not always easily distinguished without appropriate chemical tests. Before the blowpipe on charcoal bismuthinite fuses readily, boils, and spits. In the reducing flame it gives a bismuth globule, leaving a lemon-yellow bismuth-oxide coating on the charcoal. The most characteristic reaction to bismuth is the bright-red bismuth-iodide fringe around the assay when it is fused with potassium iodide (assay button). Dissolves readily in HNO_3 with separation of floating sulphur.

Genesis and Occurrence: Bismuthinite is found exclusively in *hydro-thermal* (vein and contact metasomatic) deposits, though in conditions somewhat different from those of antimonite. As an accessory it is found in deposits of tin, tungsten, arsenic, often associated with native bismuth, arsenopyrite, chalcopyrite, sometimes with native gold, topaz, beryl, pyrite, galena, and many other sulphides. It rarely forms deposits of its own.

In the oxidised zone it is easily destroyed, altering to basic carbonates in the form of pseudomorphs after bismuthinite.

In the USSR, the most important deposits are in Central Asia, e.g., *Brich-Mullavthe YG* bismuthinite, together with native bismuth occurs in a number of quartz veins in limestones, associated with pyrite arsenopyrite, chalcopyrite, etc. In a number of deposits in the Trans Baikal region bismuthinite accompanies such minerals as cassiterite, chalcopyrite, arsenopyrite, etc., and sometimes occurs in granular masses.

The world's largest bismuth deposits are in Bolivia and in Peru. Genetically these deposits are associated with young extrusives.

Uses: Bismuthinite ores are the main source of bismuth used in the production of fusible alloys, glass with a high birefringence, chemicals, medical preparations, etc. We shall also describe tetradymite here, though it belongs to a different group.

Tetradymite, Bi_2Te_2S: "Tetradymos" means fourfold in Greek (it often forms crystal fourlings). Synonym: bismuth telluride.

Chemical Composition: Bi 59.3 per cent, Te 36.2 per cent, S 4.5 per cent. Se, Au, Cu, Pb may be admixed in negligible amounts. Selenium content may be as high as 1 per cent. Gold, often paragenetically associated with tetradymite, may be present as inclusions.

System, trigonal; *symmetry,* ditrigonal scalenohedral. Lamellar or tabular aggregates are common. Tabular pr rhombohedral crystals usually occur as fourlings with {0118} and {0115} as the composition plane. Crystal structure, typically lamellar.

Colour, steel grey. *Lustre,* strongly metallic. *Hardness,* 1.5 to 2.0. Thin lamellae are flexible. *Cleavage,* {0001} excellent. Specific gravity, 7.24 to 7.54. Poor conductor of electricity thermoelec trical.

Diagnostic Features: In many external features resembles molybdenite; differs from it by stronger lustre, higher specific gravity, behaviour before the blowpipe and reactions with acids.

Unlike molybdenite, it is readily fusible on charcoal; with S and KI gives a bright orange coating (the presence of bismuth). Dissolves in concentrated H_2SO_4, and upon heating imparts to the solution the purple colouration characteristic of Te. Readily soluble in HNO_3.

Genesis being the most common mineral among tellurides, tetradymite is found most often as an accessory in hydrothermal gold deposits. Paragenetically, it associates with various sulphides: pyrrhotite, chalcopyrite, pyrite, tetrahedrite, bismuthinite, and also gold, etc.

In the oxidised zone of ore deposits, it readily decomposes, giving rise to so-called bismuth ochres.

In the USSR, it has been found in the *Frolovsky* mine of the Turya group (Northern Urals); the *Shilovo-Iset* gold deposit (66 km east of Sverdlovsk); the Kumak deposit (Southern Urals) and in a number of localities of Western and Eastern Siberia.

Elsewhere in the world it has occurred in a number of gold deposits of the USA, Mexico, British Columbia, etc.

Uses: Tetradymite forms no deposits in its own right. Being an accessory in bismuth and gold deposits it may be a source of bismuth and tellurium obtained as by-products.

PYRITE GROUP

We shall review here an extensive group of $4X_2$-type compounds where A = Fe, Co, Ni as well as Mn, Pt, and Ru, and X_2 = S_2, Se_2, As_2, AsS and SbS. These are the so-called disulphides, diarsenides, sulphoarsenides, and sulphoantimo nides. All of them have many properties in common.

This large group may be divided, according to specific mineralogical features, into four sub-groups:

(1) *Pyrite sub-group* (in the narrow sense), where FeS compound is dimorphous (pyrite and marcasite);

(2) *Cobaltite sub-group,* which comprises sulphoarsenides and sulpho-antimonides of Ni and Co (Fe in minor amounts); the members crystallise in the cubic system; although these minerals are similar to pyrite in crystal structure, their symmetry is of a lower order;

(3) *Lollingite sub-group* comprising diarsenides of Fe, Ni and Co which crystallise in the orthorhombic system;

(4) *Arsenopy rite sub-group* which includes sulphoarsenides and sulpho-antimonides, chiefly those of Fe, crystallising in the monoclinic and orthorhombic systems.

Colloidal and metacolloidal iron bisulphide occurring as finely-dispersed black masses has the special name of *melnikovite,* but it is not regarded as a separate mineral, since X-ray examination gives a debye graph of either pyrite or marcasite.

In the structure of the orthorhombic FeS_2 modification of marcasite we find the same $[S_2]^{2-}$ groups, as in pyrite. The Fe ions occupy the corners and the centre of the rhombic unit cell they are surrounded by the $[S_2]^{2-}$ ionic groups. The $[S_2]^{2-}$ pairs are shown to incline to the *c* axis to about at both ends on the centres of the Fe ionic triads, i.e., the pattern is the same as for pyrite.

From these examples of the pyrite and marcasite structures it is obvious that the same type of coordination may give rise to crystal structures of entirely different symmetry.

To mention, but a few of the physical properties distinguishing pyrite group minerals, we shall note that they are the *hardest* of all sulphides and similar compounds, their hardness ranging from 5 to 6 (sperrylite) and even as high as 7 to 8 (laurite). Another noteworthy feature is the absence of perfect cleavage. All minerals of the pyrite group are poor conductors of electricity.

Pyrite, FeS_2. "Pyros" is the Greek for fire. The name is probably associated with the property of pyrite to give off sparks on being struck, or with its strong lustre. Synonyms: fool's gold, iron pyrite.

Chemical Composition: Fe 46.6 per cent, S 53.4 per cent. Often contains minor amounts of impurities: Co (cobalt pyrite), Ni, As, Sb, sometimes Cu, Au, Ag, etc. The content of the latter elements is due to minute mechanical inclusions of foreign minerals, sometimes in a finely-dispersed state. In such instances we are in effect dealing with solid pseudosolutions or crystallosols.

Habit: Well-developed crystals are very common. Depending on the predominance of different faces, the habit of crystals may be cubic, pentagonal dodecahedral, less often octahedral. The crystals may be as large as several centimetres across. The faces show characteristic striations parallel to the (100):(210) edges, i.e., *a* : *e* (cf. 1176 and *c* and 117a). This striation conforms to the crystals structure of pyrite and is always oriented perpendicular to each adjacent face, i.e., the external elements of symmetry fully agree with the peculiar structure of pyrite. Twins are parallel to (110), rarely to (320).

Aggregates: Pyrite occurs in many rocks and ores as impregnated crystals or rounded grains. It is often found as massive aggregates. In sedimentary rocks it often occurs as globular concretions, sometimes of radial-fibrous structure, and also as secretions in the cavities of shells. Grape-like or botryoidal formations associated with other sulphides are common.

Colour, pale brass-yellow, often with yellowish-brown and iridescent tarnish; finelydispersed sooty varieties are black. Streak, brownish to greenish black. *Lustre,* strong metallic. *Hardness,* 6 to 6.5. Rather brittle. *Cleavage,* {100} and {111}, some times {110} poor. Fracture uneven, sometimes conchoidal. Specific gravity, 4.9 to 5.2. Poor conductor of electricity Thermoelectrical. Some varieties have rectifying properties.

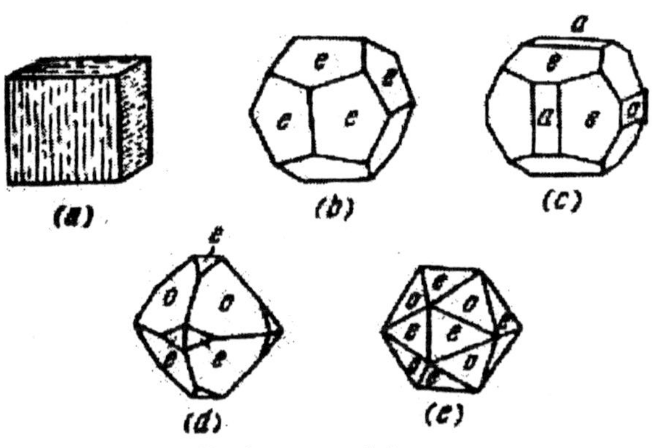

Pyrite crystal forms

Diagnostic Features: Readily identifiable by colour, crystal forms, striation, high hardness (the only widespread sulphide to scratch glass). These properties easily distinguish, it from marcasite, chalcopyrite, pyrrhotite, and millerite which are somewhat similar to it in colour.

Before the blowpipe decrepitates and fuses into a magnetic globule. Gives up readily part of its sulphur, which burns with a blue flame. In the closed tube, part of the sulphur sublimates leaving FeS monosulphide as the residue. Decomposes with difficulty in nitric acid (easily when powdered) with separation of sulphur. Does not dissolve in dilute hydrochloric acid.

Genesis and Occurrence: Pyrite is the most widespread sulphide in the earth's crust, and forms under most diverse geological conditions.

1. As tiny impregnations, it is observed in many *magmatic* rocks. It is mostly epigenetic in relation to silicates, and is associated with superimposed hydrothermal phenomena.
2. In *contact-metasomatic* deposits, it is an almost constant accessory of sulphides in the skarns and magnetite deposits. May contain cobalt. Its formation, as well as that of other sulphides, is connected with the hydrothermal stage of contact-metamorphic processes.
3. As an accessory, it is widely distributed in *hydrothermal* ore, deposits of most different composition and occurs in paragenesis with mostdiverse minerals. It is found not only in ore bodies but also in the enclosing rocks as impregnations of well-formed crystals of metasomatic origin (metacrystals).
4. It is no less widespread in *sedimentary* rocks and ores. Pyrite and marcasite concretions in sandy-argillaceous sediments, in beds of coal, iron, manganese, bauxites, etc., are well known. Its formation in these rocks and ores is believed to be associated with the decomposition of organic remains in the absence of free oxygen in the deeper parts of water reservoirs. Its most common paragenetic associates are marcasite, inelnikovite (black powdery variety of iron disulphide), siderite ($FeCO_3$), etc.

In the oxidised zone pyrite is unstable, like most sulphides, and prone to oxidise to ferrous iron sulphate, which, in the presence of free oxygen, readily passes into ferric iron sulphate.. The latter undergoes hydrolysis, i.e., decomposes into insoluble iron hydroxide (limonite) and free sulphuric acid which passes into solution. Such is the process of the formation of widespread limonite *pseudomorphs* after pyrite.

Pyrite itself often forms pseudomorphs after wood and

other organic remains; endogenous formations may contain pyrite pseudomorphs after pyrrhotite, magnetite (Fe_3O_4), haematite (Fe_2O_3), and other iron minerals. These pseudomorphs are due probably to the action of H_2S on the minerals.

Pyrite deposits are countless belonging to most diverse genetic types, although the most common are endogenous deposits.

The USSR has many huge pyrite deposits.

Uses: Pyrite ores are an essential material for sulphuric acid production. The average sulphur content of ores mined for this purpose ranges from 40 to 50 per cent. The ores are roasted in special kilns. The resulting sulphurous gas (SO_2) is oxidised to H_2SO_4 by nitrogen oxides in the presence of water vapour.

Marcasite, FeS_2. The name comes from the ancient Arabic term for pyrite and marcasite.

Chemical composition. Fe 46.6 per cent, S 53.4 per cent. Impurities: minor amounts of As, Sb, Tl, etc. orthorhombic.

Habit: More seldom short columnar or lance-shaped. Often twinned resembling cockscombs. Occurs as concretions and also as sinters, botryoidal, reniform, crusts, and other irregular forms. Common as pseudomorphs after organic remains.

Colour, brass yellow with a greyish or greenish tinge. *Streak,* dark, greenish grey. *Lustre,* metallic. Elsewhere in the world, it has occurred in a number of gold deposits of the USA, Mexico, British Columbia, etc.

Uses: Tetradymite forms no deposits in its own right. Being an accessory in bismuth and gold deposits it may be a source of bismuth and tellurium obtained as by-products.

MOLYBDENITE GROUP

This group includes sulphur compounds of tetravalent Mo and W ions. Tungsten sulphide is an exceptionally rare mineral.

Hardness, 5 to 6. Brittle. Cleavage, {101} imperfect. Specific gravity, 4.6 to 4.9 (lower than that of pyrite). Poor conductor of electricity.

Diagnostic Features: Crystals of marcasite are distinguished from those of pyrite by their lance-shaped and tabular forms. Hard to distinguish from pyrite in concretions or compact masses. Fresh fracture, in contrast to pyrite, shows a greenish hue. Under the microscope in polished sections, it is readily distinguished from pyrite by optical anisotropy. Debyegraphs also differ sharply from those of pyrite. Before the blowpipe and in acids behaves exactly as pyrite.

Genesis and Occurrence: Much less abundant than pyrite. Occurs both in endogenous and exogenous deposits. Endogenous marcasite occurs inhydrothermal, mostly vein, deposits. It is formed, as a rule, in the latest stages of mineralisation. Most commonly found in drusy cavities as small crystals on cavity walls, sometimes as powdery coatings on crystals of quartz, calcite, galena, sphalerite, grey ores, and other minerals, less often as crusts and sinters.

In sedimentary, chiefly carboniferous sandy-argillaceous rocks, marcasite occurs as concretions, irregular grains, pseudomorphs after organic remains, and also as a finely-dispersed black sooty substance (melnikovite). Marcasite is frequently confused with pyrite.

In the oxidised zone, marcasite decomposes more readily than pyrite into iron sulphates and free sulphuric acid; oxygen deficiency results in the evolution of native sulphur. The final product of marcasite oxidation is iron hydroxide (limonite).

It has been shown experimentally that, as different from pyrite, marcasite forms relatively easily from acid solutions. Unlike pyrite, marcasite does not occur in large massive ore deposits.

A notable hydrothermal sulphide deposit containing marcasite, along with pyrite, in appreciable amounts is the Blyavinskoye deposit in the Orenburg Region, Southern Urals, where marcasite occurs as fine crystalline *sporadically* distributed aggregates. Marcasite associates not only with pyrite but also with sphalerite, wurtzite, chalcopyrite, quartz, etc.

Marcasite-bearing sedimentary rocks are widespread in the Soviet Union. Among them are the carboniferous sediments of sandy-argillaceous at Borovichi, Novgorod Region, in which marcasite concretions of various shapes are found. Polished sections of the concretions point to epigenetic formation of marcasite which only replaced the cementing material between the sand grains, the latter preserving their lamellar arrangement in the mass of marcasite. Besides marcasite, some sections of the concretions may contain pyrite, sometimes concentrically arranged. Some concretions have a coating of galena on their surface, deposited in exactly the same way, i.e., by replacing the cementing material in the sand masses. Generally speaking, however, the association of iron sulphides with galena in concretions is a rare phenomenon.

The argillaceous sedimentary deposits on the eastern slope of the Middle Urals, east of Sverdlovsk at *Kurya-Kamensk* and *Troitsk-Bai-novo*, are well known for the wide variety of marcasite concretion forms found in them. Besides spheric nodules, there are many reniform and radial concretions terminating in well-bounded lance-shaped crystals.

Hydrothermal deposits from which well-formed marcasite crystals have been recorded are known in Germany.

Uses: Like pyrite, when found in huge massive deposits, marcasite may be a material for sulphuric acid production.

Sperrylite, $PtAs_2$. Named after the chemist who discovered the mineral in Sudbury ores.

Chemical Composition: Pt 56.5 per cent, As 43.5 per cent. Impurities (per cent): Rh (up to 1.6), Fe (up to 0.4), Cu (up to 0.7), Sb (up to 0.6), sometimes Sn (up to 3.6).

Occurs almost exclusively in crystals, mostly small ones. Habit, cubic, octahedral, more rarely pentagonal dodecahedral. Often in these combinations of forms: {100}, {111}, {HO}, {201}, etc. Twins are rare.

Colour, tin white. *Streak,* dark grey. *Lustre,* strong, metallic. *Hardness,* 6 to 7 (in the class of sulphides and arsenides it is second to laurite in hardness). *Cleavage,* cubic. Specific gravity, 10.5 to 10.7 (highest for this class of minerals). Poor conductor of electricity.

Diagnostic Features: The most important diagnostic features are colour, high hardness, high specific gravity, and occurrence in crystals, high acid-resistance, and platinum and arsenic reactions.

Before the blowpipe on charcoal fuses readily into a white metallic globule with a spongy surface, and gives off white vapour of arsenic oxide. On a red-hot platinum tablet, fine grains fuse instantaneously with separation of As_2O_3; the sponge platinum is welded to the tablet. Insoluble in acids, even in aqua regia.

Genesis and Occurrence: Occurs in deposits of copper-nickel sulphide ores of the Sudbury type, in genetic association with basic igneous rocks (gabbro-norites and gabbro-diabases). Paragenetic associates are pyrrhotite, chalcopyrite, and pentlandite. The most common associate among the platinum-group minerals is palladium platinum.

It has been found in the same paragenetic associations in the pegmatite formations of basic rtiagmas in the Bushveld complex, South Africa, where largest (up to 1.85 cm across) crystals have been found.

In the same locality, it was observed in peculiar metasomatic deposits in limestones at the contact with basic rocks of the Bushveld complex, in association with skarn minerals.

Owing to its chemical inertness, sperrylite does not decompose in the oxidised zone, and upon the weathering of the deposits passes into placers, often retaining its crystal faces. In the USSR, sperrylite occurs in river placers of *the Zeya* and *Timpton* districts of the Eastern Siberia, Chita Region, commonly in well-formed small crystals.

Uses: Has definite economic value as a platinum-rich

mineral. Even if its content in an ore is negligible, it may be extracted as a by-product. An example is Sudbury where sperrylite is extracted along with other platinum minerals from the slimes of copper-nickel ore concentration.

Cobaltite, CoAsS. Synonym: cobalt glance. An iron-rich variety is *ferrocobaltite.*

Chemical Composition: Co 35.4 per cent, As 45.3 per cent, S 19.3 per cent. According to analyses, Co content ranges from 26 to 34 per cent, As from 42 to 48 per cent and S from 18 to 21 per cent. Ni (2 to 3 per cent) and Fe (up to 8 per cent and sometimes, as in the case of ferrocobaltite, up to *124.* Cobal-16 per cent may also be present.

System, cubic; *symmetry,* pentagonal tritetrahedral. Crystal structure, very much like that of pyrite Octahedron.

Habit: Octahedral, cubic, and pentagonal-dodecahedral (cf. 117). Hence the most common forms are the combinations characteristic of pyrite, especially {111} and {210} shown in 124. Twins along (110) and (111) are rare. Cobaltite is also found as irregular grains and massive.

Colour, white to steel grey with a reddish tinge. Iron-rich varieties are dark grey or greyish black. *Streak,* grey black. *Lustre,* metallic. *Hardness,* 5 to 6. Brittle. *Cleavage,* distinct, cubic. *Specific gravity,* 6.0 to 6.5. Poor conductor of electricity.

Diagnostic Features: On close examination is readily identifiable by the reddish tinge, high hardness, and often by typical combinations of forms {100}, {111}, and {210}. Is distinguished from linneite, which is of similar colour, by higher hardness. Weathered specimens are characteristically associated with intensely pink erythrite.

Before the blowpipe "fuses into a slightly magnetic globule, giving a coating of As_2O_3 on charcoal. Imparts deep blue colour (cobalt reaction) to the borax bead. Decomposes in nitric acid with separation of S and As_2O_3 (pink solution).

Genesis and Occurrence: Occurs chiefly as a typical mineral of hydrothermal processes in contact-metasomatic and vein deposits. Commonly associated with the arsenic-sulphur minerals of cobalt and iron, and also with chalcopyrite, sphalerite, quartz, skarn minerals, ferruginous chlorite, tourmaline, apatite, etc. On weathering, cobaltite, as other arsenious cobalt compounds, alters readily to earthy or crystalline erythrite whose pink colour is very conspicuous in the oxidised zones of sulphide-arsenic deposits of cobalt.

In the USSR, cobaltite occurs in the *Dashkesan* iron ore deposit of contact-metasomatic origin (Kirovabad district, Azerbaijan) in hydrothermally altered skarns in the hanging wall of a magnetite ore body. Its paragenetic associates include chalcopyrite, pyrite, sphalerite, molybdenite, magnetite, garnet, calcite, quartz, apatite, etc.

Large deposits of cobaltite occur at *Cobalt*, Ontario, Canada, associated with safflorite, skutterudite, smaltine, chloanthite, niccolite, gersdorffite, native silver, argentite, dolomite, calcite, and other minerals. Other notable deposits are at *Skutterud*, Norway, and at *Tuna-berg*, Sweden.

Uses: Gobaltite is one of the principal sources of cobalt in commercial ores. In view of the high prices fetched by cobalt, sulphide-arsenic ores may be worth mining, even when their cobalt content is as low as 0.1 to 0.2 per cent.

Cobalt has a number of valuable properties which determine its uses: (1) its various compounds make stable deep-blue and green pigments used in glass staining and tinting of ceramics; this was already known deep in antiquity; (2) when used as an alloying component in steels, it imparts to them great hardness and heat resistance, and also outstanding magnetic properties; (3) gives a number of important alloys with Cr, Mo, and W, etc.

Gersdorffite, NiAsS. Synonym: nickel glance. Cobalt- and iron-rich varieties are known.

Chemical Composition: Ni 35.4 per cent, As 45.3 per cent, S 19.3 per cent. Ni content usually varies from 26 to 40 per cent,

As 37 to 56 per cent, S 6 to 19 per cent. Common *impurities* are Co, Fe, Sb, etc.

System, cubic; *symmetry*, pentagonal tritetrahedral. Crystal structure, similar to that of pyrite.

Habit: Octahedral or cubic. Most common forms are: {100}, {111}, {110}, {210}, {311}. Rare twins are along (111). Occurs mostly in granular aggregates.

Colour, silver white to steel grey. *Streak*, grey-black. *Lustre*, metallic. *Hardness*, 5.5. Brittle. *Cleavage*, sometimes parallel to (111). *Specific gravity*, 5.6 to 6.2. Good conductor of electricity.

Diagnostic Features: Hardly distinguishable superficially from a number of arsenious minerals, such as smaltite ($CoAs_2S_3$), chloanthite ($NiAs_{2-3}$), ullmannite (NiSbS), arsenopyrite (FeAsS), etc.; hence the need for microscopic examination and chemical tests for Ni, As, and S; and in the presence of the isomorphs of Fe, Co, Sb, also for quantitative determinations of at least the principal elements.

Before the blowpipe on charcoal fuses into a globule which gives the nickel reaction. Decomposes in nitric acid with separation of S and As_2O_3. The solution is green, indigo.

Genesis and Occurrence: Gersdorffite is a mineral occurring predominantly in *hydrothermal* deposits. It is associated paragenetically with arsenides and sulphides of nickel, such as niccolite, millerite, chloanthite, rammelsbergite, ullmannite, etc. Other possible associates are various sulphides, and also carbonates (calcite, ankerite, dolomite), and quartz. In the oxidised zone, gersdorffite, as other nickel arsenides, turns into bright green annabergite cating the presence of $Ni.Ni_3[AsO_4]_2.8H_2O$).

In the USSR, gersdorffite has been found in very few localities and in negligible quantities. At the *Berezovskoye* gold mines in the Urals has been observed in small grains in coarse-crystalline dolomite, and also in the *Berikul* gold deposit, Western Siberia, in association with rammelsbergite, niccolite, etc.

The largest finds have been made at *Hartz* (Germany), the *Erzge-birge mountains* (Saxony) and at some other localities.

Uses: Not very important commercially, since it usually occurs only as an accessory in the sulphide-arsenious ores of nickel and cobalt.

Lollingite, $FeAs_2$. Named after the town of Lolling in Carinthia, Austria. Was first described by Mohs.

Chemical Composition: Fe 27.2 per cent, As 72.8 per cent. The Fe:As ratio is somewhat variable. Often contains minor amounts of S (up to 6 per cent) and Sb (up to 5 per cent). Certain varieties are enriched in cobalt (glaucopyrite), which suggests a continuous series of solid solutions: lollingite-safflorite ($FeAs_2$—$GoAs_2$). In addition to cobalt, nickel is usually present in minor amounts.

System, orthorhombic; *symmetry,* rhombic dipyramidal. Crystal structure, similar to that of marcasite, though in the type of coordination of As around Fe, it differs somewhat from both marcasite and arsenopyrite. According to V.I. Mikheyev, the structure is retained when iron is replaced by cobalt. With increasing (Co + Ni): Fe ratio, the size of the unit cell gradually increases along the *c* axis, while the length of the other two edges of the cell remains about the same.

Habit: Predominantly prismatic 125. Also occurs massive. *Colour,* silver-white to steel-grey. *Streak,* grey-black. *Lustre,* metallic. *Hardness,* 5 to 5.5. Brittle. *Cleavage,* sometimes along {010} and {110} distinct. Fracture uneven. *Specific gravity,* 7.0 to 7.4 (much higher than that of arsenopyrite). Other properties. Good conductor of electricity.

Diagnostic Features: Externally very similar to arsenopyrite for which it is often mistaken. The minerals differ most sharply in specific gravity. Some differences are also observed in polished sections under the microscope.

Is less fusible in the blowpipe flame than arsenopyrite. In the closed glass tube, yields only metallic arsenic, unless it contains a considerable admixture of sulphur. Dissolves in nitric acid with separation of As_2O_3.

Genesis and Occurrence: Is more rare in nature than arsenopyrite, and usually is found in minor amounts. Confined exclusively to *hydro-thermal* vein and metasomatic deposits, often associated with arsenopyrite, iron and copper sulphides, Co arsenides, as well as with calcite, siderite, quartz, etc.

In the oxidised zone, decomposes to scorodite (Fe''' [AsO_4]-$2H_2O$). In the USSR, lollingite was found in the Caucasus, Urals, Central Asia, Western and Eastern Siberia. These the most notable is the *Aguyurm* deposit in Eastern Karategin on the left bank of the Aguyurm river, Central Asia, where it is common in arsenopyrite ores. In the *Sokhondinskoye* tin ore deposits in Kyrin district, Chita Region, the mineral had occurred as continuous masses and veinlets up to 3 cm thick. A characteristic feature is that it forms later than arsenopyrite and overgrows crystals of the latter.

Occurrences of lollingite in Carinthia, Hartz, Saxony, Norway, Canada, and at other places have been mentioned in literature.

Uses: Constitutes an arsenic ore, being an iron arsenide with the richest arsenic content.

Arsenopyrite, FeAsS. Synonyms: arsenical pyrites, arsenic iron. Variety: *danaite* which is a cobalt-bearing arsenopyrite; cobalt-rich modifications have the general name of *glaucodot.*

Chemical Composition: Fe 34.3 per cent, As 46.0 per cent, S 19.7 per cent. Chemical analyses show frequent deviations from these figures, especially with regard to As and S. Common impurity is Co, more seldom Ni and Sb. In quite a few deposits arsenopyrite is gold-bearing, the gold being present as microinclusions, but for the most part as finely-dispersed phase, i.e., arsenopyrite in this case is in effect a crystallosol.

System, monoclinic; *symmetry,* prismatic. Crystal structure. Although morphologically arsenopyrite belongs to the marcasite series, X-ray analysis shows that its ideal structure is monoclinic, and appears orthorhombic only due to twinning. Each Fe ion is surrounded inside the distorted octahedron by three S ions, three apexes of the octahedron being occupied by S and the other three by As. The As and S ions occupy the apexes of the distorted tetrahedron: the As ions are surrounded by three Fe ions and one S ion, and the S ions by three Fe ions and one As ion.

Habit: Very often occurs in excellent, usually prismatic, short-columnar to rod-like and acicular crystals. Also common are pseudodipyramidal crystals formed by the uniform development of prisms of the first and second order. Most common forms are {101}, {230}, {210}, {140}, etc. Characteristically striated faces parallel to the *c* axis. Crystals commonly occur in drusy cavities, but are also frequent in the wall rocks of deposits as metasomatically formed metacrystals. Rather frequent twins, often cruciform.

Aggregates: Granular and rod-like when in continuous masses.

Colour, tin-white (crystal faces) to steel-grey (fracture). Often with yellow tarnish. *Streak,* grey-black, sometimes with a brownish tinge. *Lustre,* metallic. *Hardness,* 5.5 to 6. Brittle. *Cleavage,* rather distinct along {101} and also along {001}. *Specific gravity,* 5.9 to 6.2. Conductor of electricity. The temperature of decomposition ranges from 430 to 675°, the bond between Fe and As being weaker than that between Fe and S.

Diagnostic Features: Identifiable by tin-white crystal faces, high hardness, and the presence of iron, arsenic, and sulphur as principal components. When struck with a hammer, gives off the odour of garlic.

Crystal forms are also very characteristic. From lollingite ($FeAs_2$) is distinguished by lower specific gravity. From arsenious compounds of nickel and cobalt (smaltine and chloanthite) is distinguished in granular masses with certainty only by qualitative chemical testing and microscopic examination of polished sections with microchemical tests.

Fuses in the reducing flame emitting garlic odour; gives a magnetic globule with tombac-brown fracture. In the closed tube, gives a red sublimate of arsenic sulphide and later a black ring of metallic arsenic. Decomposes in nitric acid with separation of S and As_2O_3.

Genesis and occurrence. Arsenopyrite is a mineral of *hydrothermal* origin and one of the most common arsenic bearers in endogenous deposits.

In typical hydrothermal, vein and metasomatically formed deposits, it segregates mostly during the high-temperature stages of mineralisation.

Not infrequently forms deposits in its own right, being the principal ore mineral. It also occurs, as an accessory in most diverse tin, tungsten, bismuth, copper, lead, zinc, and other deposits. Its most common non-metallic associates are quartz, tourmaline, feldspars, micas, carbonates, sometimes beryl, topaz, etc.

In the course of oxidation in the weathering zone arsenopyrite decomposes rather rapidly to form scorodite (Fe-"$[AsO_4]2H_2O$), usually as pale yellowish and greenish loose earthy masses (when mixed with iron hydroxides, it turns brown).

In the USSR, tens of deposits are known in which arsenopyrite is the chief ore-forming mineral. Of the numerous deposits outside the USSR mention should be made of the great deposit at *Boliden* Sweden) where arsenopyrite bears much gold which is not recovered completely by mechanical concentration.

Uses: Arsenopyrite ores are the principal source of various arsenic compounds used partly in pest control, and also in the manufacture of dyes, leather, and in other chemical industries.

The minimum arsenic content making mining worthwhile is commonly accepted at 5 to 6 per cent. In complex concentration of polymetallic ores arsenic may be obtained from its minerals as a by-product, especially from the smelter fumes.

Skutterudite Group

This group includes arsenides of Ni and Co of AX3 composition, in which isomorphs with a high Fe content are present in considerable quantities. All of them crystallise in the cubic system.

X-ray examinations show that these minerals differ fundamentally from those of the foregoing group in crystal structure, though their composition may often be expressed in terms of diarsenides.

A characteristic feature of the skutterudite crystal structure is the formation by arsenic atoms of "fourfold As_4 groups which occupy the apexes of a *square.*

These flat groups are located in the middle of the edges and faces, parallel to the faces of the unit cube. Each Co ion is surrounded by 6 As ions.

The crystal structures of smaltite and chloanthite proved to be absolutely identical with that of skutterudite, despite their lower arsenic content; only their unit cells are slightly larger than those of skut-terudite.

It is small wonder, therefore, that these minerals have strikingly similar physical and chemical properties. To avoid repetition, we shall give below a general description of the group.

Smaltite, CoAsx where $x = 0$ to 0.5. Synonym: smaltine, grey cobalt. Smalta is a deep-blue cobalt pigment.

Chloanthite, NiAs, where $x = 0.5$ to 1.0. "Chloantes" is the Greek for green-colouring. The name was probably given because of the green colouration imparted to acid solutions by nickel compounds or because the oxidation products (annabergite) of the mineral are green, in contrast to cobalt arsenides which yield pink secondary mineral (erythrite).

Chemical composition: Chemical composition is v· riable. The content of individual constituents varies over a wid range.

Chemical Composition of the Minerals of Skutterudite Group in Weight Percentages

Elements	*Skutterudite*	*Smaltite*	*Chloanthite*
As	76.4 to 72.9	71.6 to 63.4	73.5 to 70.1
S	0 to 1.7	0 to 1.4	0 to 0.6
Co	20.5 to 10.8	24.1 to 13.8	3.6 to 6.3
Ni	0 to 9.4	1.0 to 15.0	21. 2 to 14. 5
Fe	0 to 5.8	1.2 to 7.3	2.8 to 5.2

Chemical analyses also reveal the presence of Cu, Bi, sometimes Pb and Ag, which most probably are mechanical impurities.

System, cubic; *symmetry,* didodecahedral. The size of the unit cell increases with increasing Ni and Fe content. $a_0 = 8.189$ (for Skutterudite), 8.24 (for smaltite), 8.28 (for chloanthite).

Habit: Cubic, cubic octahedral, or octahedral. Occur as dendrites or massive granular aggregates.

Colour, tin-white, sometimes with a grey or iridescent tarnish. *Streak,* grey-black. *Lustre,* metallic.

Hardness, 5.5 to 6. Brittle. *Cleavage,* {100}, {111} distinct. Fracture often conchoidal. *Specific gravity,* 6.4 to 6.8. Conductors of electricity.

Diagnostic Features: The minerals are very difficult to identify by external features, especially when they are intimately intergrown with other similar arsenides of nickel, cobalt, and iron. When massive, they resemble externally arsenopyrite, lollingite, gersdorffite, ullmannite, safflorite, and rammelsbergite. May be positively identified only upon microscopic examination of polished sections, and chemical and X-ray analyses.

Before the blowpipe they give a magnetic globule and emit a strong arsenic odour. Cobalt-rich varieties decompose in nitric acid imparting a pink coloration to the solution upon heating, whereas nickel-rich varieties give it a yellow-green colour.

Genesis and Occurrence: All these minerals occur in

paragenetic association with other arsenides of cobalt and nickel only in *hydrothermal* deposits of the Schneeberg type.

On weathering Skutterudite and smaltite decompose to erythrite ($Co_2[AsO_4]2'8H_2O$) which appears as a pink-coloured selvage, while the weathering product of nickel Skutterudite and chloanthite is bright-green annabergite ($Ni_3[AsO_4]_2.8H_2O$).

Skutterudite has been found along with other cobalt minerals in the arsenic-cobalt deposit of *Ak-Jilga,* the Alai ridge (Southern Kirghizia) and at *Nizhni-Seymchan* in the northeast of the Asian part of the USSR, where it associates with glaucodot, cobaltite, arseno-pyrite, pyrrhotite, and other minerals in quartz veins.

Notable localities in other countries are *Skutterud,* Norway, where skutterudite was first discovered in association with cobaltite and sphene in mineralised gneisses; *Cobalt,* Ontario (Canada), where, besides skutterudite and nickel skutterudite, chloanthite and smaltine have been found associated with other minerals of cobalt, nickel, and silver; Erzgebirge mountains at *Jachymov* (Czechoslovakia), *Schneeberg, Anna-berg* (Saxony).

Uses: These minerals sometimes occur together with other arsenides and sulphoarsenides of nickel and cobalt in large amounts, in which case they are of definite economic importance.

Sulphosalts

This class comprises complex compounds, chemically similar to salt-like compounds, and known as sulphosalts. Like oxysalts, they are sub-divided into sulphobases and sulphoanhydrites, i.e., cations and complex anions.

Although the minerals of this class are many and varied, their constituents are few.

In the vast majority of cases, the sulphoanhydrites in sulphosalts are As_2S_3, Sb_2S_3, and Bi_2S_3 which are designated accordingly as sulphoarsenites, sulphoantimonites, and sulphobismuthites.

It should be emphasised that all the three types of sulphosalts have the same metals in their bases: Cu, Ag, and Pb. In other words, the sulphosalts of copper, silver, and lead are the most common in nature, Independent sulphosalts of Tl, Hg, and Fe are extremely rare; only Zn and Mn are found as isomorphous admixtures in any appreciable amounts.

Sulphovanadates, sulphoarsenates, and sulphoantimon ates of copper, i.e., sulphosalts having V_2S_5, As_2S_5 and Sb_2S_5 (with pentavalenl ions) for sulphoanhydrites are also found in negligible amounts. Less clear is the problem of the existence of the so-called sulphostannates and sulphogermanates (with Sn_2S_3 and CeS_2).

The vast variety of sulphosalts is largely due to the different ratios between sulphobases and sulphoanhydrites in compounds of the same qualitative types.

Crystal structures of sulphosalts, similarly to oxysalts, differ from simple sulphurous compounds in that their crystal lattices always in elude structural units in the form of anionic compact groups such a $[SbS_4]^{5-}$ and $[SbS_3]^{3-}$, etc. Indeed, the simplest cubic and trigonal crystal structures of sulphosalts include anionic groups in the form of obtuse trihedral pyramids (analogous to $[SO_3]^{2-}$ complex anions in sulphites) $[AsS_3]^{3-}$, $[SbS_3]^{3-}$, where As and Sb are trivalent. Examples of these most simple structures are those of proustite (Ag_3AsS_3), pyrargyrite (Ag_3SbS_3), tetrahedrite (Cu_8SbS_3), etc. Further more, the structures may include tetrahedral groups such as $[VS]^{3-}$, $[AsS_4]^{3-}$, $[SbS_4]^{3-}$, in which V, As and Sb are pentavalent. Complex sulphoanions are less compact than complex oxyanions, this being possibly due to the lower polarising capacity of As^{3}+, Sb^{3}+, Bi^{3}+, V^{5}+, As^{5}+ and other cations.

Sulphosalts have somewhat specific physical properties. As compared with simple sulphurous compounds, sulphosalts are generally less hard (especially, the sulphosalts crystallising *in the* lower systems, i.e., the majority of mineral sulphosalts), poorer reflectivity and lower resistance to acids. These peculiar physical properties are undoubtedly related in some way to the specific crystal structures of these compounds.

The best way to classify sulphosalts would be to divide them into larger groups according to the metals in their sulphobases, followed by sub-division according to the composition of sulphoanhydrites. Then the classification of sulphosalts will be as follows:

A. *Copper sulphosalts,* i.e., compounds of $7iCu_2S_2X_2S_3$, where X = As^3+, Sb^3+, and Bi^3+ and those of the $3Cu_2S$-X_2S_5-type, where *X* = V^5+, As^{5+}, and Sb^5. This class would thus include sulphoarsenites, sulphoantimonites, and sulphobismuthites of copper, as well as sulphovanadates, sulphoarsenates, and sulphoantimonates of copper.

B. *Silver sulphosalts:* $rcAg_2S_2X_2S_3$, where *X* = As^{3+}, Sb^{34}", and Bi^3+, i.e., sulphoarsenites, sulphoantimonites, and sulphobismuthites of silver.

C. *Lead sulphosalts:* nPbS, X_2S_3 where X = As^3+, Sb^{3+}, and Bi^3+, i.e., sulphoarsenites, sulphoantimonites, and sulphobismuthites of lead.

Tetrahedrite Group

The minerals included under this heading comprise a large isomorphous group of so-called fahlite (grey ores) with a common chemical formula A_3XS_3 where A_2 = Cu_2, less often Ag_2, Cu, Zn, Fe, rarely Hg; and X = As, Sb, and rarely Bi (in negligible amounts).

X-ray analyses indicate that the formula $A_{12}X_4S_{13}$ would be the more precise. This agrees well with a certain excess of sulphur shown by precise chemical analyses.

Depending on the prevalent sulphoanhydrite in these compounds, the following mineral species are distinguished: tennantite ($Cu_{12}As_4S_{13}$) and tetrahedrite ($Cu_{12}Sb_4S_{13}$). The so-called mixed grey ores of the Gu_{12} (As, Sb)$_4S_{13}$ composition are the most common in nature.

The crystal structure of grey ores is rather complex and generally resembles that of chalcopyrite or sphalerite, except for the size of the unit cell being twice as large.

All the mineral species and varieties included in this group according to their chemical composition have many physical properties in common and hence are described below.

Tennantite,$Cu_{12}As_4S_{13}$or$3Gu_2S$-As_2S_3. Named after the chemist Tennant.

Tetrahedbite, $Cu_{12}Sb_4S_{13}$ or $3Cu_2S$-Sb_2S_3. The name derives from the shape of the crystals in which it occurs and which is generally common to the grey ores of different composition.

Chemical Compositio: The content of individual elements in different grey ores varies as follows (per cent):

Cu .. .22 to 53 Hg ... 0 to 17.0 As .. .0 to 20.0

Ag... 0 to 18 Ni... 0 to 3.5 Sb .. .0 to 29.2

Zn ... 0 to 9 Go ... 0 to 4.2 Bi.. .0 to 4.5 (13.07)

Fe...0 to 13 Mn...0 to 1.5 S .. .20.6 to 29.1

Varieties according to chemical composition are *freibergite,* a silver-rich tetrahedrite; *zandbergite,* a zinc-rich tennantite or tetrahedrite; *ferrotetrahedrite* (up to 13.08 per cent of iron); *ferrotennantite,* etc.

System, cubic; symmetry, hexatetrahedral. In the tennantite-tetrahedrite isomorphous series the size of the unit cell increases as arsenic is replaced by antimony, and copper by silver. Crystals found in cavities are tetrahedral with this combination of forms: {111}, {110}, {112}, {100}, etc. Commonly in continuous masses or impregnations of irregular grains. Often twinned along (111), more seldom along (100).

Crystal structure. Grey ores have half of the copper ions in fourfold coordination (the same as sphalerite), and the other half in threefold coordination (i.e., without the fourth apex). The As (or Sb) cations are surrounded by three S ions each.

Colour, steel grey to iron black (Fe-rich varieties). *Streak* has the same colour, sometimes with a brownish and even cherry-red tinge (tennantite). Opaque. *Lustre,* metallic or sub-metallic. *Hardness,* 3 to 4. Brittle. *Cleavage,* practically none. *Specific gravity,* 4.4 to 5.4. Arsenious varieties have lower *specific gravity* than antimonious varieties. Poor conductor of electricity.

Diagnostic Features: Readily identifiable by grey fracture and obvious brittleness (when scratched with a knife, gives a fine powder and no shiny trace is left, as in the case of chalcocite and argentite, that are otherwise rather similar to tetrahedrite). In colour and brittleness, is similar to bournonite ($CuPbSbS_3$), which has a lower hardness and a somewhat stronger lustre.

Before the blowpipe on charcoal fuses readily into a grey globule giving off As_2O_3 and Sb_2O_3. The globule gives the copper and often the iron reaction. Decomposes in nitric acid with separation of S and Sb_2O_3. Behaviour with reagents depends on composition.

Genesis and Occurrence: Grey ores, tetrahedrite in particular, are rather widespread in *hydrothermal* copper deposits of different types. They are present in minor amounts in ores of most widely varying composition. Their most frequent paragenetic associate is chalcopyrite, less often sphalerite, galena, pyrite, arsenopyrite, bournonite, and other minerals.

On weathering the ores decompose readily, yielding various alteration products: covellite, malachite, azurite, limonite; the arsenic present in the ores gives rise to scorodite ($Fe[AsO_4].2H_2O$); while antimony goes to form various oxides and hydroxides.

Grey ores are widespread in the USSR They are found in all copper and lead-zinc deposits, though rarely in large amounts. The richest grey-ore deposits are those known as *Blagpdatniye mines.* In these deposits the grey ores are strongly enriched with gold, and also contain antimony and arsenic.

They are associated mainly with pyrite, chalcopyrite, and to some extent with galena. Well-formed tetrahedrite crystals are found in drusy cavities at the *Berezovskoye* gold mines. Tennantite crystals have been recorded from hollow fissures in many pyrite deposits of the Middle Urals.

Uses: Grey ores never form large deposits in their own right. Together with other copper-bearing sulphurous compounds in commercial deposits, they are a source of copper. In the smelting of copper ores containing tennantite, arsenic volatilises in the form of As_2O_3, a harmful impurity, in the smelter fumes. The same happens in the smelting of ores containing arsenopyrite, enargite, and other arsenious compounds as impurities. At large plants such gaseous arsenic is extracted from the sublimation products which are thus rendered harmless; this may yield a considerable quantity of arsenious material as a by-product.

Enargite Group

This group includes compounds of Cu_3XS_4 or $3Cu_4S$-X_2S_5 type, where X = V, As, and Sb, i.e., pentavalent elements.

Enargite, Cu_3AsS_4. "Enargis" means distinct in Greek (probably for the distinct cleavage of the mineral).

Chemical Composition: Cu 48.3°/$_{p}$, As 19.1 per cent, S 32.6 per cent. Impurities: Sb (up to 6.5 per cent), Fe (up to 5.7 per cent), in negligible amounts also Pb, Zn, and Ag (possibly in the form of inclusions of foreign minerals).

System, orthorhombic; *symmetry,* rhombic dipyramidal. Crystals mostly columnar, vertically striated, more rarely tabular; consist of combinations of forms {110}, {001}, {100}, {010}, etc. Usually massive granular or disseminated. Crystal structure, similar to that of wurtzite and is regarded as pseudohexagonal.

Colour, steel grey to iron black. *Streak,* grey-black. Opaque. *Lustre,* sub-metallic, strong. *Hardness,* 3.5. Brittle. *Cleavage,* {110} perfect, also {010} rather distinct. *Specific gravity,* 4.4 to 4.5. Other properties. Poor conductor of electricity.

Diagnostic Features: In external features is very similar to black sphalerite, from which it differs by cleavage which is perfect only in one direction, and by stronger lustre.

Fuses in the blowpipe flame on charcoal giving a coating of As_2O_3. When roasted with soda, gives a copper globule. Dissolves in nitric acid with separation of floating sulphur.

Genesis and occurrence. Sometimes occurs in quantity in *hydrothermal* deposits in association with grey ores, chalcopyrite, galena, pyrite, and other minerals.

Quite common are tennantite pseudomorphs after enargite called "green enargite". A comparison of the chemical formulas of these minerals shows that under endogenous conditions the process is limited to As_2S_5 altering to As_2S_3.

In the oxidised zone of deposits decomposes readily yielding malachite, azurite, olivenite (copper arsenate), and other secondary minerals.

In the USSR, enargite occurs as an admixture to other minerals. Important commercial deposits of enargite are at *Butte,* Montana, USA; *Chuquicamata,* Chile; South-West Africa, and elsewhere.

Uses. When found in quantity, enargite is an ore of copper and arsenic.

Bournonite Group

This includes binary sulphosalts of copper and lead of $CuPbXS_3$-type, such as bournonite and aikinite. *Bournonite,* $CuPbSbS_3$. Synonym: berthonite.

Chemical Composition. Cu 13.0 per cent, Pb 42.5 per cent, Sb 24.7 per cent, S 19.8 per cent. Impurities: Fe (up to 5 per cent), Ag (up to 3 per cent), traces of Zn and Mn.

System, orthorhombic; *symmetry,* rhombic dipyramidal. Well-formed crystals are found only in cavities and often are pseudotetragonal and thick-tabular with developed faces of forms {001}, {010}, {100}, {Oil}, {110}, etc. Frequently twinned along (110). Recurrent cruciform or wheel-like twins. Usually occurs in irregularly-shaped grains or massive. Crystal structure, as yet undeciphered.

Colour, steel grey to lead grey, often with tombac-brown tarnish (on crystal faces). *Streak,* grey. Opaque. *Lustre,* metallic. *Hardness,* 2.5 to 3. Brittle. *Cleavage,* {010} imperfect. Specific gravity, 5.7 to 5.9. Non-conductor of electricity. Diagnostic features. Externally has some resemblance to grey ores, except for stronger lustre.

Before the blowpipe on charcoal with soda fuses readily into a black bead which upon removal of all lead (by careful ignition with boric acid) yields *with difficulty* a copper globule. The black bead on roasting with KI gives a yellow-green coating of PbI_2. Dissolves in nitric acid with separation of S and Sb_2O_3.

Genesis and Occurrence: Occurs in *hydrothermal* lead-antimony ore deposits. Generally is closely associated with tetrahedrite and galena, sometimes is found as selvages on the boundary between them.

In the oxidised zone of deposits decomposes readily, sometimes altering to malachite, cerussite ($PbCO_3$), and antimony oxides.

In the USSR, bournonite occurs in appreciable amounts in the quartz veins of the *Nagolny Ridge,* the Ukraine, in association with galena, grey ores, boulangerite, jamesonite, etc., also at the *Darasun* gold deposit (Eastern Trans-Baikal Region) in grey ores, galena, etc..

Outside the USSR, occurs in large amounts at *Pribram,* Czechoslovakia; *Clausthal* and *Andreasberg,* Harz, Germany, and in many deposits in the USA, Mexico, Peru, Chile, and elsewhere.

Uses: When found in quantity has economic value as an ore of lead and copper.

Aikinite, $CuPbBiS_3$. Was first discovered in the Urals near Sverdlovsk early in the last century. Synonym: patrinite.

Chemical Composition. Cu 11.0 per cent, Pb 36.0 per cent, Bi 36.2 per cent, S 16.8 per cent. Impurities: Te and Au (the latter as inclusions).

System, orthorhombic; *symmetry,* rhombic dipyramidal. Often found in acicular and rod-like crystals in quartz, sometimes massive. Faces often vertically striated.

Colour, lead grey to steel grey, often with dark-brownish or brownish tarnish. *Streak,* grey-black, shiny. Opaque. *Lustre,* metallic. *Hardness,* 2 to 2.5. Brittle. *Cleavage,* {010} imperfect. *Specific gravity,* 6.1 to 6.7.

Diagnostic Features. Characteristic acicular or rod-like crystals, but without chemical tests for Bi, Pb and Cu the mineral is hard to identify.

Fuses readily in the blowpipe flame, gives off fumes and leaves a white and yellow coating on charcoal. Gives a metallic bead which in turn yields with difficulty a copper globule when roasted with boric acid. Gives the reaction for Bi with KI. Dissolves in nitric acid with separation of $PbSO_4$ and sulphur.

Genesis and Occurrence: It is a rare mineral occurring in quartz veins of *hydrothermal* origin in association with pyrite, chalcopyrite, grey ores, galena, arsenopyrite, native gold, sometimes scheelite, wolframite, and other minerals.

Well-known are the rod-like and acicular crystals of aikinite in the transparent or translucent crystals of quartz occurring at the *Berezov-skoye* gold deposit, northeast of Sverdlovsk. It is often paragenetically associated with galena, grey ores, pyrite, chalcopyrite, and especially with native gold.

In the oxidised zone it is unstable. Decomposes into earthy yellow or green-yellow masses of the so-called bismuth and lead ochres.

Uses: In itself it is of purely mineralogical interest. At the Berezovsk ore deposit its presence in veins is a good guide to gold.

Proustite Group

The group comprises sulphoarsenites and sulphoantimonites of silver of the Ag_2XS_3 type where X = As and Sb. Proustite and pyrargyrite are the most common minerals of this group. Although both minerals have analogous chemical formulas and crystallise in the same symmetry, chemical analyses show that they do not form a continuous series of isomorphous mixtures. As has been shown by experiments, these compounds can be intermixed infinitely only at high temperatures.

We shall also describe here the rare sulphosalts of silver—stephanite and polybasite. Pboustite, Ag_3AsS_3. Named after Proust, the chemist, who was the first to recognise two separate ruby-silver ores: arsenical and antimonial.

Chemical Composition: Ag 65.4 per cent, As 15.2 per cent, S 19.4 per cent. Analyses show that Ag content ranges from 63.4 to 67.6 per cent, As from 12.3 to 20.2 per cent, and S from 13.1 to 20.2 per cent.

System, trigonal; *symmetry,* ditrigonal pyramidal. Crystal structure. The AsS_3 groups occupy every corner and the centre of the rhombohedral cell.

Each such group is a low pyramid with as an apex. The apexes of all the pyramids are oriented parallel to the threefold axis.

Habit: Commonly occurs in scalenohedral crystals with acute rhombohedrons 131). The principal forms are hexagonal prism {1120} and ditrigonal pyramids {2131}, {1124}, etc. Faces are striated obliquely. Occurs mostly as disseminated grains, sometimes massive.

Colour, scarlet-vermilion (like cinnabar). *Streak,* vermilion. Translucent. Lustre, strong adamantine. For Li-light: $n_y = 3.088$, $n_z = 2.792$. *Hardness,* 2 to 2.5. Brittle. *Cleavage,* {1011} distinct. Fracture conchoidal. *Specific gravity,* 5.57 to 5.64. Non-conductor of electricity. Diagnostic features. Externally can hardly be distinguished from pyrargyrite. Usually is somewhat lighter in colour than the latter. Crystals have fewer faces than crystals of pyrargyrite. However, the main difference between the two is in the As and Sb contents.

Other minerals similar to proustite in colour and lustre are miargyrite ($AgSbS_2$), cinnabar, cuprite (Cu_2O), and zincite (ZnO). Miargyrite, if not in crystals, can be distinguished from proustite only by the As: Sb ratio, which is determined by chemical analysis. Cinnabar is easily identified by its behaviour before the blowpipe (volatilises completely). Cuprite is identified by its octahedral crystals, dark brownish-red streak, and also paragenetic association with native copper and other copper minerals. Zincite has an orange-yellow streak and high hardness (4 to 4.5).

Before the blowpipe fuses readily, gives off the arsenic odour, and yields a coating of As_2O_3 and Sb_2O_3 (when antimony is present). When roasted in the reducing flame on addition of soda, its bead yields silver. Dissolves in nitric acid with separation of S and As_2O_3 (in the case of pyrargyrite, Sb_2O_3 is separated).

Genesis and Occurrence: Is widely distributed in *hydrothermal* veins of lead-zinc-silver ores (of the so-called noble quartz-calcite formation). Occurs with minerals formed in the latest stages of hydrothermal processes, such as pyrargyrite. Common paragenetic associates of these minerals are galena, sometimes native silver, and also sulphoarsenites and sulphoantimonites of lead, silver, and copper of different composition. In some deposits they are associated with arsenides of nickel and cobalt.

In the oxidised zone proustite and pyrargyrite decompose, sometimes forming native silver and argentite. But usually silver, in the unstable form of Ag_2SO_4, is capable of migrating, especially in the presence of free sulphuric acid and ferric sulphate. In a number of instances, this gives rise to silver enrichment of the lower parts of the oxidised zone.

In the lead-zinc deposits of the USSR, proustite and

pyrargyrite are mostly found as minute segregations visible in polished sections under the microscope. They occur most frequently in a number of lead-zinc-silver deposits of the *Western Verkhoyanye:* Bezymyanskoye, Verkhne-Endybalskoye, Berezinskoye, and other deposits.

The richest deposits of these two minerals are in Mexico (Zacatecas,. Guanajuato, etc.), Chile, Peru, Bolivia, and elsewhere.

Uses: Proustite and pyrargyrite are the most widespread silver-bearing minerals, and therefore are important ores of silver.

Pybargyrite, Ag_3SbS_3. From the Greek words "pyros"-fire, and "argyros" — silver. Synonyms: dark ruby-silver ore 132). Many physical properties of this mineral are similar to those of proustite. We shall mention those that are peculiar to pyrargyrite. Colour in reflected light is dark red to iron black; the corners of crystals and fragments are translucent. Streak, dark cherry-red. Specific -gravity, 5.77 to 5.86 (higher than that of proustite).

Stephanite, Ag_5SbS_4 or $5Ag_5S$-Sb_2S_3. Ag *o* {0001}, *r* {1011}, content 68.5 per cent. System, orthorhombic. Occurs in *m* {1010}, *y* {3251}. short-prismatic crystals and massive. Colour, grey-black. Streak, black. Lustre, metallic. Hardness, 2 to 2.5. Brittle. Cleavage, {010} distinct. Specific gravity, 6.2 to 6.3. Fuses and decrepitates in the blowpipe flame on charcoal, giving a coating of Sb_2O_3. With soda yields a silver globule. It is decomposed by dilute nitric acid with separation of S and Sb_2S_3. Occurs together with other silver minerals, usually in very small amounts, in veins of hydrothermal origin. Has been recorded from deposits of Saxony and Harz, Germany; Mexico, and elsewhere.

Polybasite, $(Ag, Cu)_{16}Sb_2S_n$ or $8(Ag, Cu)_2S$-Sb_2S_3. Ag content 62.1 to 74.9 per cent, Cu 3 to 10 per cent. System, monoclinic. Occurs in tabular or short-prismatic crystals of pseudohexa gonal habit. *Colour,* grey-black. *Streak,* black with a reddish tinge. *Lustre,* metallic. *Hardness,* 2 to 3. *Cleavage,* parallel to {001}. *Specific gravity,* 6.27 to 6.33. Fuses readily in the blowpipe flame on charcoal into a metallic globule, giving off fumes, and produces a coating of Sb_2S_3. Fused with a phosphate, gives a greenish-blue bead (copper). Occurs together with other silver sulphosalts in low-temperature hydrothermal veins in several localities: Jachymov and Pfibram, Czechoslovakia; Schemnitz, Hungary; Zacatecas, Guanajuato and Dur-ango, Mexico, and elsewhere.

Lead Sulphosalts

The compounds included in this group are mostly sulphoarsenites, sulphoantimonites, and sulphobismuthites of lead. All of them behave differently from the sulphosalts of copper and silver, only in a few instances forming binary compounds with them. The chemical constitution of many of them has not yet been finally established. We shall review only two minerals: boulangerite and jamesonite. *Boulangerite,* $Pb_5Sb_4S_n$ *or* $5PbS'2Sb_2S_3$.

Chemical composition: Pb 55.4 per cent, Sb 25.7 per cent, S 18.9 per cent. Pb content varies from 54 to 58 per cent, part of it being due to the presence of galena as a mechanical impurity. Sometimes contains up to 1 per cent of copper.

System, monoclinic; *symmetry,* prismatic. Crystals very rare. Usually occurs in fine-granular or felted aggregates.

Colour, lead grey to iron black. *Streak,* grey-black with a brownish tinge. Opaque. *Lustre,* metallic. *Hardness,* 2.5 to 3. Brittle. *Cleavage,* {100} distinct. *Specific gravity,* 6.23.

Diagnostic Features: Identifiable usually by fine-fibrous aggregates and brownish streak. Cannot be distinguished positively from a number of other, rarer, sulphoantimonites of lead, without chemical and X-ray analyses. Fuses readily before the blowpipe. On charcoal with soda yields a lead globule and a thick white coating of Sb_2O_3 near the assay. Dissolves in nitric acid and hot hydrochloric acid.

Genesis and Occurrence: Occurs in *hydrothermal* deposits of lead-zinc ores together with copper and lead sulphoantimo nites, galena, sphalerite, pyrite, arsenopyrite, etc.

In the oxidised zone decomposes readily, forming cerussite ($PbCO_3$) and antimony hydroxides. In the USSR, occurs in the *Nagolny Ridge*, the Ukraine, associated with sphalerite, galena, bournonite, grey ore, and other minerals; also found in several deposits in the Nerchinsk district (Eastern Trans-Baikal region): the *Algachinskoye* (in large amounts), *Klichkinskoye, Darasunskoye,* etc.

Uses: When found in large masses, is an ore of lead.

Jamesonite, Pb4FeSb$_6$S$_{14}$ or 4PbS'FeS-3Sb$_2$S$_3$. A variety which does not contain iron is called *plumosite.*

Chemical Composition: Chemical composition does not always agree exactly with the formula. Pb content 40 to 50 per cent, Fe up to 10 per cent, Sb about 30 per cent, S about 20 per cent. Frequent impurities are copper, zinc, and silver.

System, monoclinic. Often occurs in acicular and capillary crystals, sometimes as ingrowths in crystals of quartz, sphalerite, and other late-crystallising minerals in drusy cavities.

Colour, lead grey. *Streak,* grey-black. Faces sometimes show a bluish-grey tarnish. *Lustre,* metallic. *Hardness,* 2 to 3. Brittle. Fracture uneven. *Cleavage,* {001} distinct. *Specific gravity,* 5.5 to 6.0.

Diagnostic Features: Indistinguishable to the naked eye from other sulphoantimonites of lead occurring in acicular forms. Chemical and X-ray analyses are necessary for positive identification. Fusible in the blowpipe flame. On charcoal yields a lead globule. Roasted with soda, gives a white coating of Sb_2O_3. Dissolves readily in nitric acid. Genesis and occurrence. Occurs relatively rarely in *hydrothermal* deposits of lead-zinc ores in association with quartz, galena, sometimes sphalerite, and other minerals in drusy cavities. In the USSR, has been recorded from the deposits of *Nagolny Ridge,* the Ukraine, in association with calcite, at *Darasunskoye* (Trans-Baikal region). A notable locality is Zimapan, Mexico, where it has been found in quantity.

Beginning with this type of compounds we shall deal with minerals sharply differing in properties from any we have reviewed before. These are predominantly compounds with typical ionic bonding, imparting altogether different properties to the minerals. Their most typical representatives are halides of metals.

	H^{1}										
H^{2}	Li^{3}	Be^{4}	B^{5}	c^{6}	N^{7}	O^{8}					F^{9}
N.^{10}	Na^{11}	Mg^{12}	Al^{13}	Si^{14}	P^{15}	s^{16}					Cl^{17}
Ar^{18}	K^{19}	Ca^{20}	Sc^{21}	Tl^{22}	v^{23}	Cr^{24}	Mu^{25}	Fe^{26}	Co^{27}	Ni^{28}	Ci^{29}
Zn^{30}	Ga^{31}	Ge^{32}	As^{33}	Se^{34}	Br^{35}						
Kr^{36}	Rb^{37}	Sr^{38}	Y^{39}	Zr^{40}	Nb^{41}	Mo^{42}	To^{43}	Ru^{44}	Rh^{45}	Pd^{46}	Ag^{47}
Cd^{48}	In^{49}	Sn^{50}	Sb^{51}	Te^{52}	I^{53}						
Xe^{54}	Cs^{55}	Ba^{56}	TR$^{57\text{-}71*}$	Hf^{72}	Ta^{73}	w^{74}	Re^{75}	Os^{76}	Ir^{77}	Pt^{78}	Au^{79}
Hg^{80}	*Tl*^{81}	Pb^{82}	Bi^{83}	Po^{84}	At^{85}						
Rn^{86}	Fr^{87}	Ra^{88}	Ac^{89}	Th^{90}	Pa^{91}	U^{92}					

From the chemical standpoint, the minerals concerned are salts of hydrofluoric, hydrochloric, hydrobromic, and hydroiodic acids. Accordingly, the minerals are classified as *fluorides, chlorides, bromides* and *iodides*. Moreover, there exist hydrous salts and more complex compounds which contain additional oxigenated anions such as $[OH]^{1-}$, O^{2-}, rarely $[SO_4]^{2-}$ and $[IO]^{1-}$. These are the so-called oxihalide compounds transitional to typical oxygen compounds.

The principal halide-forming elements, occupy the left-hand half of the periodic table, mostly groups I and II. Halide compounds of heavy metals are on the contrary almost of no importance in the mineralogy of natural formations and are formed under special conditions.

The specific crystallochemical features of halides. Crystal structures have been studied exhaustively only for some simple anhydrous compounds AX and AX_2 types which crystallise in the cubic system.

Thus, it has been found that the halides of light metals possess structures with a typical *heteropolar* (ionic) bonding, whereas halides of heavy metals whose cations are strongly polarising develop *homopolar* (covalent) bonding (or bonds of a transitional type) between ions. These peculiarities in the structure of minerals determine their physical properties.

Since halides with a typical ionic bonding contain weakly charged light-metal cations with big ionic radii and, therefore, of low polarising capacity, it is but natural that these minerals are transparent, colourless (occasional colouring is, as a rule, allochromatic), have low specific gravity, high water solubility, low refractive indices, and consequently a weak vitreous lustre.

On the other hand, in the case of cations of heavy metals with an 18-electron outer shell (e.g., Cu, Ag), which tend to polarise the surrounding anions rather strongly and to form crystal structures with a covalent type of bonding, the properties of the halides will differ substantially: high specific gravity; the fact that a number of compounds possess, though weak, but idiochromatic colouring; materially higher refractive indices; much lower solubility.

Now we shall consider the properties dependent on the principal halide anions such as F^{-1}, Cl^{-1}, Br^{-1}-, and I^{-1}.

First, it should be noted that the size of the F^{-1} anion is materially different from that of the other anions, as will be seen from the comparison of the lengths of their radii (angstroms): 1.33 1.81 1.96 2.19

This difference has much to do with the selection of compound-forming cations (in conformity with coordination numbers and types of crystal structure), with the stability of the compounds, and hence with their physical and chemical properties.

No wonder, therefore, that the bulk of fluorine in the earth's crust is combined with Ca and partly with Al and Si (in binary compounds), whereas chlorine and the far less common bromine and iodine are combined chiefly with Na, K, (Rb), (Cs), and Mg (in hydrous salts). As different from the chlorides, bromides, and iodides of heavy metals (Au, Ag, and

Hg), their fluorides do not occur in nature. Chlorides of light metals are highly soluble in aqueous media, whereas their fluorides are for the most part immune to the action of water.

The melting and boiling points of fluorides are incomparably higher than those of the chlorides of the same metals. For instance, the boiling point of SnF_4 is 705°, whereas that on $SnCl_4$ is 114°; that of AlF_3 is 1000°, and that of $AlCl_3$ only 81°, etc.

Specific features of the behaviour of halides in nature. Most interesting are the specific features of the behaviour of F, Cl, Br, and I halides in the course of geological processes.

Solubility of Halides in Water at 18°C

(moles per litre of saturated solution)

Anions	*Cations*	*Li*	*Na*	*K*	*Mg*	*Ca*	*Sr*	*Ba*	*Pb++*
	F	0.11	1.06	12.4	0.02	0.03	0.001	0.03	0.003
	Cl	13.1	5.42	3.9	5.1	5.4	3.0	1.7	0.05
Br	12.6	6.9	4.6	4.6	5.2	3.4	2.9	0.02	
I	8.5	8.1	6.0	4.1	4.8	3.9	3.8		

Magmatic processes do not provide conditions for these elements to concentrate in any large amounts. Fluorine and chlorine enter into the composition of a number of minerals, mainly silicates and phosphates, only as additional anionic constituents (mostly in pegmatites and contact-metasomatic formations). The bulk of these elements, evidently as volatile metal compounds, passes into hydrothermal solutions. That chlorine and fluorine are a part of the volatile magmatic constituents is additionally confirmed by the emanation of HCl and HF, sometimes abundant, in the gaseous products of volcanic eruptions. Thus, in 1919 in the Valley of Ten Thousand Smokes, Alaska, as estimated, 1,250,000 tons of gaseous HCl and 200,000 tons of HF emanated together with water vapours.

Many hydrothermal formations abound in such halides as fluorite (CaF_2) and contain some Al fluorides; metal chlorides are not found there, except for very rare microscopic crystals in the droplets of solutions occluded in some minerals (*quartz* and galena).

Under exogenous conditions, on the other hand, chlorides of Na and, to a lesser extent, those of K, Mg, and other metals are formed, frequently in enormous masses, in land-locked bodies of salt water together with sulphates, borates, and other water-soluble compounds. In this respect, bromides and iodides stand next to chlorides. Today 70 to 75 per cent of the total chlorine (and probably bromine) and over 90 per cent of iodine in the earth's crust are dissolved in oceanic waters.

Contrary to this, fluorides are practically absent from salt-bearing sediments. Oceans and seas are gigantic collectors of dissolved chlorides. At the same time the fluorine content of sea water is as low as 0.8 g per cu m. It has been established that this element is partly assimilated by higher organisms entering into the composition of their skeletons, especially the dental enamel which is composed almost exclusively of calcium fluoride. The migration of fluorine under exogenous conditions has another peculiarity. In the process of rock and ore weathering, much fluorine is liberated along with chlorine. However, having a very strong affinity with calcium, fluorine, on its way to the sea, in the main precipitates from the solutions in the form of poorly soluble CaF_2 compound which is retained in continental sediments. That explains the insignificant fluorine content in sea water.

Classification of Halides: Thus, the properties of halides as well as their geochemical role in the mineral formation processes allow their division into two classes:

Class 1: Fluorides.

Class 2: Chlorides, bromides, and iodides.

FLUORIDES

Fluorides as minerals occur rather sparsely in nature, though as many as 15 elements are capable of combining with fluorine. The most important fluoride is that of calcium, CaF_2, which occurs as an independent compound. Much less important are B, Al, and Si. Other elements enter into the composition of some very rare fluorides.

The most widely distributed fluorides recognised so far are confined mostly to hydrothermal formations, whereas the rare fluorides are found in sublimation products of volcanic activity. Judging by paragenesis of minerals they must be formed at relatively high temperatures. Only CaF_2 in newly-formed disseminated crystals occurs frequently in the oxidation zones of ore deposits and in certain sedimentary rocks.

We shall review here fluorite and cryolite — two of the natural compounds of the fluoride group.

Fluorite, CaF_2. "Fluorum" is the Latin name of the element F. Synonym: fluorspar. Like other fluorine-rich minerals, it is a good flux for ores, accelerating smelting.

Chemical Composition: Ca 51.2 per cent, F 48.8 per cent. Occasionally contains Cl (mostly yellow varieties) as an isomorphous impurity or bituminous substances with a characteristic odour. Among other impurities are Fe_2O_3, rare earths, in rare instances uranium (up to a few per cent), fluorine, and helium.

System, cubic; *symmetry,* hexoctahedral. Crystal structure is typical of many *AX* type compounds. It has two coordination numbers: 8 for Ca and 4 for F. The F^1-ions occupy the corners, while the Ca^{2+} ions occupy the centres of all the smaller cubes.

Habit: In cavities may occur as well-formed cubic crystals, less often as octahedral and dodecahedral crystals. Besides {100}, {111}, and {110} forms, {210}, {421}, etc., forms may sometimes be present. The cube faces are usually smooth, while the octahedral ones appear frosted. Sometimes, the cube faces carry a parquet-like pattern 17. Twinning frequent along (111).

Aggregates: Mostly found as impregnations or granular masses, less often as earthy masses (ratofkite).

Colour, rarely colourless or water-transparent; commonly

yellow, green, blue, violet to violet-black. Curiously, the colour disappears on heating and reappears on exposure to X-rays. In colourless crystals, a violet colour may be induced by the vapours of metallic calcium and electric discharges. This suggests that in a number of cases the colouring is caused by the appearance of electrically neutral Ca and F atoms in the crystal structure. *Lustre*, vitreous, $n = 1.434$.

Hardness, 4. Brittle. When subjected to prolonged unidirectional pressure, shows plastic deformations. *Cleavage*, perfect octahedral not rhombododecahedral, contrary to the concept that cohesion should be the weakest between the planar nets that are farthest from each other. This is explained by the fact that among the (111) planar nets each net of Ca ions is interlaid by two nets of similarly charged F ions, and hence it is between these that cohesion is the weaker. Specific gravity, 3.18 (for impure varieties ranges from 3.0 to 3.2). Other properties. Often displays fluorescence (the very term originated from the name of this mineral). In cathode rays, fluorite usually gives a violet luminescence with a peculiar bluish-green hue. Fluorescence also appears on heating (thermoluminescence).

Diagnostic Features: With practice, is rather easily identified by crystal forms, octahedral cleavage, weak and somewhat dull vitreous lustre, and hardness.

Before the blowpipe decrepitates, fluoresces; the edges of fragments are rounded with difficulty (at 1270°). On complete evolution of all fluorine, an infusible calcium oxide (CaO) residue remains. Water solubility is very poor. Decomposes completely only in strong sulphuric acid with separation of HF. Much less soluble in nitric and hydrochloric acids.

Genesis and Occurrence: Forms mainly in the course of *hydrothermal* processes, being often an accessory of ore mineral in veins. May occur in association with most diverse minerals of hydrothermal origin.

Found also in some *sedimentary* rocks, but does not form

la:ge masses with a high F content. As a compound poorly soluble in water, CaF_2 is among the first to precipitate, sometimes in the amorphous state from salt solutions. No wonder, therefore, that the rare fluorite accumulations are confined to early chemical sediments, e.g., to deposits of gypsum, anhydrite, calcite, and dolomite. As rare neogenic formations, it occurs in the oxidised zone of ore deposits, e.g., as minute crystals on goethite sinters.

As an accessory, fluorite occurs in numerous deposits of non-ferrous and rare metals. In the USSR, fluorite is the principal mineral in the *Kalanguy* deposit (Trans-Baikal region), where it occurs as a thick breccia vein in sandstones and schists and is composed of concentrically-zonal and rod-like aggregates of various tints (white, yellow, and reddish-yellow). Earthy fluorite of sedimentary origin (ratofkite) has been found in dolomitised limestones on the banks of the Ratofka River near the town of Vereya. (Moscow Region), on the right bank of the Osuga River (Kalinin Region), and elsewhere.

Uses: The bulk of fluorite (about 70 per cent) is used in steal making as a flux. The chemical industry uses fluorite as a basic ingredient of a number of fluoric compounds, such as fluoric acid for etching glass, for obtaining hydrogen peroxide from sodium peroxide, etc.; artificially produced cryolite (Na_3AlF_6) is used for obtaining electrolytically metallic aluminium from alumina, also for preparation of enamels and glazes. Transparent and colourless crystals are used in optics for making microscope lenses without spherical and chromatic aberrations.

Cryolite, Na_3AlF_6. "Cryos" is the Greek for ice, "lithcs"-stone. Has been evidently named for ice which it closely resembles in lustre and refractive index.

Chemical composition: Al 12.8 per cent, Na 32.8 per cent, F 54.4 per cent. Sometimes contains Fe— as an impurity.

System, monoclinic, *symmetry,* prismatic. Pseudocubic; at 500° becomes cubic.

Habit: Cube-like with best developed faces {001} and {010}. Commonly occurs in masses comprised of large individuals.

Colour, Colourless but more often grey-white, yellowish or reddish, rarely black. *Lustre,* vitreous. Tig =1.34.

Hardness, 2 to 3. Brittle. No *cleavage. Specific gravity,* 2.95 to 3.01. On slight crystal. Diagnostic features. Characteristic crystal forms and weak vitreous lustre. Accurate identification possible only by blowpipe analysis and by measuring optical constants.

Readily fuses before the blowpipe, and even in the flame of a candle imparting a reddish-yellow coloration to the flame and yielding a transparent bead which turns into white enamel on cooling. Gives the fluorine reaction in the glass tube. On prolonged roasting on charcoal gives an alumina deposit which turns blue when moistened with cobalt nitrate. Completely soluble in strong nitric acid.

Genesis: Occurs in pegmatites in which it is formed from residual solutions enriched with fluorine. An important economic deposit is at *Ivigtut,* Western Greenland, where it occurs in granite that has been converted to greisen and has the shape of a large vertical stock extending to a great depth. Cryolite is found there in coarse-crystalline masses together with other complex fluorides, rare sulphides, cassiterite, etc. In the *Ilmen Mountains* (Southern Urals) cryolite has been recorded from a topaz working and was accompanied by chiolite, amazonite, and other minerals.

Uses: Cryolite is usually obtained artificially, and used in the metallurgy of aluminium, in the manufacture of milk-white glass, enamel for kitchenware, and for other purposes.

Chlorides, Bromides, and Iodides

Chlorides, in contrast to fluorides, are widely distributed in nature. Chlorine compounds are known for the following 16 elements: H (as HCl in gaseous volcanic emanations), N (in HN_4), Na, Mg, Al, K, Ca, (Rb), (Cs), Fe, (Ni), Cu, Ag, Hg, Pb, and Bi.

The most important of these are Na and Mg chlorides which under exogenous conditions often form, together with other soluble salts, thick salt deposits of sedimentary origin. The heavy metal (Cu, Ag, and Pb) compounds are of secondary importance. The rest of the elements form rarely occurring minerals.

It should be noted that under exogenous conditions potassium and sodium play different parts geochemically. In marine basins where concentration of chlorous salts takes place, the NaCl content is 3.5 per cent, while that of KCl is hardly 0.6 per cent to 0.7 per cent, although their clarks (average content) in the lithosphere are roughly the same and upon weathering of rocks these elements pass into solutions in roughly equal amounts. Chemical analyses of soils and continental deposits show that while the bulk of sodium in the surface and river waters reaches lacustrine and marine basins, potassium is, to a considerable degree, absorbed in the upper parts of the weathering crust (soils, argillaceous rocks) and assimilated by plants. As known, the ash of plants is always rich in potassium. Hence it is clear why sodium compounds are sharply predominant over potassium compounds in the crystallisation products of drying salt-water basins.

Bromides are known only for Ag and are found very seldom in the oxidised zones of silver-bearing sulphide deposits in arid and hot climate. The bulk of bromide is disseminated as an isomorphous impurity in widespread chlorides of light metals, especially in bishofite ($MgCl_2 \cdot 6H_2O$), together with which it is accumulated in land-locked lacustrine and marine basins, mainly in the residual brines.

The bromine content of sea water is 0.008 per cent. Sometimes is found in large quantities in mineral springs.

Iodides are also rare, but occur in more mineral species. Iodides exist for Ag, Cu, and Hg, i.e., metals whose ions are strongly polarising. They occur in the same conditions as bromides. Although chlorides of Na, K, and Mg vary widely in iodine content, it remains negligible in all of them. In larger quantities, it is present in the products of life activity of algae from the ash of which it is extracted. It is also found in waters of oil reservoirs and in volcanic muds. The very high solubility of iodine salts limits their accumulations to regions of arid climate. Such is, for instance, the great saltpeter ($NaNO_3$) deposit in the Atacama desert, Chile, where iodine is present in the form of iodates, its content being 0.05 per cent. Thus, chlorides are the most abundant of the halides.

Two groups of minerals of this class will be discussed: the halite group and the cerargyrite group.

HALITE GROUP

Included in this group are chlorides of univalent metals Na and K, as well as the binary hydrous chloride of K and Mg.

Halite, NaCl. "Halos" is the Greek for sea, salt. Synonyms: rock salt (when it occurs in coarse-crystalline masses in rocks) and solar salt (as friable crystalline aggregates at the bottom of bodies of salt water).

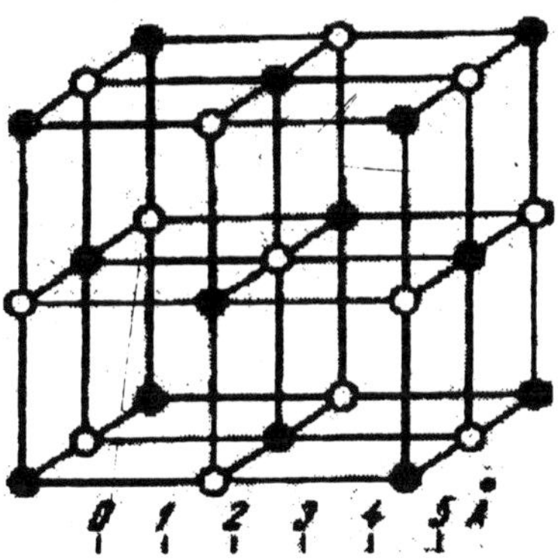

Crystal structure of NaCl.

Chemical Composition: Na 39.4 per cent, Cl 60.6 per cent. Impurities are mostly mechanical such as droplets of brine, gas bubbles, and also inclusions of clayey and organic matter, gypsum, KCl, $MgCl_2$ etc.,

System, cubic; *symmetry,* hexoctahedral. Crystal structure is characterised by the closest cubic packing of the Cl^{1-} anions, with the Na^1 cations filling all the octahedral spaces. The Na^{1+} and Cl^{1-} ions alternate at the corners of the small cubes. Coordination number of the two ions is 6, i.e., each negative chlorine ion is surrounded by six positively charged sodium ions, and vice versa.

Habit: As a rule cubic. Faces (111) and {110} occur very seldom (when crystallised from solutions of complex composition). Twins are rare, always along (111). Solar salt crystals sometimes hopper-shaped with stepped depressions on (100) faces.

Aggregates: Solar salt occurs as loose or compact crystalline-granular crusts or beds at the bottom of enclosed basins, and also as druses of crystals, which are sometimes very large.

It is noteworthy that when the water evaporates rapidly and the weather is quiet, hopper crystals float at the surface of salt brines and their upturned base keeps constantly growing. The boats are white, owing probably to occluded microscopic air bubbles. When the water is disturbed, it fills the boats which settle to the bottom and continue to grow, forming normal transparent crystals. Frequently, however, they retain inside milk-white portions with a herring-bone or envelope-like pattern.

Rock salt developing by cumulative crystallisation in the course of metamorphism forms unusually coarse-crystalline masses, as evident from the size of cleavage planes. Characteristically, recrystallisation may yield perfectly transparent masses of halite (the occluded liquids and gases evidently being "expelled" towards the perimeter of the crystal grains).

Colour, Pure masses of halite are transparent and colourless or white, but some pigments often produce different colours: grey (usually clayey particles), yellow (iron hydroxides), red (red haematite), dark brown to black (organic matter which disappears on heating), etc. Rock salt has sometimes a very peculiar intense blue colouring distributed in spots and bands, especially in those portions which have suffered considerable deformation. Such colour is obtained artificially by exposing rock salt to X-rays, or by impregnating it with metallic sodium vapours, especially after the specimen has been subjected to unidirectional compression. On heating to 200°, the colour disappears and the salt becomes colourless. The colouring is attributed to the fact that the Na^{1+} ions acquire free electrons (particles of radioactive radiation) and thus become electrically neutral atoms.

The occurrence of blue salt in those portions where potassium-containing chlorides are also present suggests that K and its usual associate Rb—both capable of emitting particles (electrons)—could cause the blue colouring. *Lustre,* vitreous, greasy on the surface of slightly weathered varies ties, *n* 1.544 (much higher than that of sylvite). *Hardness,* 2. Brittle. Prone to plastic deformation upon continued unidirectional pressure. *Cleavage,* excellent cubic (cf. 23), which agrees well with the crystal structure of the mineral. Specific gravity, 2.1 to 2.2. Poor conductor of electricity, but extremely good conductor of heat. Highly soluble in water: up to 35 per cent at room temperature (solubility changes but slightly as temperature rises the first few tens of degrees). Taste saline. Hygroscopic, but much less so than potassium and magnesium chlorides.

Diagnostic Features: Readily identifiable by low hardness, excellent cleavage, solubility in water, and salty taste. Easily fusible before the blowpipe on charcoal (at 800°), usually decrepitates and colours the flame yellow. On addition of $AgNO_3$ to a solution slightly acidified with nitric acid, gives a white flocculent AgCl precipitate.

Genesis and Occurrence: The bulk of halite, as well as of other water-soluble salts, is formed in the course of *exogenous* processes through the drying of salt lakes or shallow lagoons and bays cut off from the sea, by sand bars, in conditions of hot arid climate. In summer in such shallow bays, the evaporation of water over a vast area will cause the concentration of salts in solution to increase continuously over that in the main body of water. The resulting drop of water level will cause more water to flow from the main body adding more salt to the solution.

With the advent of cold weather, different salts are precipitated from the saturated solutions, depending on the concentration of the constituents. If a basin is cut off from the sea entirely and has no fresh water tributaries, salt may precipitate even in summer, when the solution becomes supersaturated due to evaporation.

The composition of water in inland saline lakes varies widely. This depends not only on the composition of inflowing water, i.e., of surrounding rocks, but also on the stage of the drying of the lake and on the corresponding relative concentration of constituents in the brine. In many such lakes NaCl is precipitated along with other salts.

In rock salt deposits formed in the past geologic ages halite occurs in continuous masses. In the course of orogenic movements, the salt beds emplaced in sedimentary rocks change their shape readily in view of their great proneness to plastic deformation. In this way there are often formed thick domes of complicated shape and structure, which sometimes protrude through the overlying sedimentary rocks. In this process, thin lamellar salt deposits frequently undergo very complex crumpling.

The deserts abound in salt efflorescences, which practically always contain NaCl. They disappear after periods of rain and reappear in dry weather.

Finally, halite occurs together with other chlorides as a sublimation product on the walls of volcanic craters and in the cracks in lava flows. With very few exceptions, such accumulations are small. It should be noted that in this case halite usually contains a great deal of KCl, which, as has been shown experimentally, at high temperatures enters into it in the solid-solution state.

The largest reserves of halite are concentrated in rock-salt deposits. Formed in different geologic epochs, the largest deposits, however, are confined to sediments of the Permian period notable for its prolonged hot continental climate (on the territories of Europe, Southern Asia, America, etc.).

In the USSR, the most important halite deposits are the following: the *Slavyansk-Artemovsk* at Bakhmut, the Ukraine, where 18 rock-salt strata are interbedded with clays, anhydrites, and limestones; the *Iletskaya Zashchita* where a huge rock-salt stock is overlain with only a thin blanket of drift, and rock-salt strata alternating with anhydrites are badly crumpled; *Solikamsk* (in the upper reaches of the Kama River)—the world's largest deposit of potassium and magnesium salts, enclosing also thick beds of rock salt, gypsum, etc.

The well-known *Lake Baskunchak* is an important deposit of solar salt of recent formation. The lake is located at the northern slope of a salt dome buried under a gypsum series. The crystalline salt deposits lie beneath the brine of the lake.

Notable foreign localities are *Wieliczka* rock salt deposit southeast of Krakow (Poland)—one of the oldest known, the working of which started as early as the XI century; a huge rock-salt stock near *Suez;* enormous deposits of rock salt are known in *Northern India,* along the Himalayas, in Punjab, etc.

Uses: Halite has a great variety of uses. Besides being an essential element of the diet and food-preserving agent, it is extensively used in the chemical industry for manufacture of hydrochloric acid, chlorine, bicarbonate of soda, caustic soda and a number of salts. Moreover, it is a raw material for metallic sodium employed in the production of anti-friction alloys (sodium-calcium babbits); in the production of sodium peroxide used as a bleach in the textile industry; as a catalyst in the synthesis of complex organic compounds; as a reducing agent and desulphuriser in ferrous and non-ferrous metallurgy; as an absorber of moisture and oxygen in the purification of noble gases (helium, neon, and argon, etc.); in electrical engineering in sodium-vapour filled discharge lamps, and in electrical cables with sodium "cores" sheathed with copper, etc.

Sylvite, KGl.

Chemical Composition: K 52.5 per cent, Cl 47.5 per cent. Often contains occluded liquids and gases, chiefly nitrogen, less often carbon dioxide, hydrogen, methane, and, what is most interesting, helium. Frequent mechanical impurities are NaCl and Fe_2O_3. Almost always present as isomorphous admixtures are KBr (up to 0.1 per cent) and negligible amounts of RbCl and CsCl.

System, orthorhombic; *symmetry,* hexoctahedral. Crystal structure, analogous to that of halite: face-centred cube (cf. 138). The unit cell is much larger than that of halite. Habit, cubic. Occurs fairly often in cubes truncated by octahedral faces. Twins along (111) are frequent. Aggregates. Usually granular massive, sometimes with laminated texture.

Colour. Pure varieties are water-transparent and colourless. May be milk-white due to minute occluded gas bubbles. Bright-red and pink sylvites are also crystallosols containing minutest scales of haematite (Fe_2O_3) as a coarsely-dispersed phase which precipitates on dissolution. Lustre, vitreous, n = 1.490 (lower than for halite).

Hardness, 1.5 to 2. Brittle. Shows signs of plasticity under continued unidirectional pressure. *Cleavage,* {100} excellent. Specific gravity, 1.97 to 1.99 (lower than that of halite). Taste, bitterly salty. Heat conductivity high. Readily soluble in water. Hygroscopic. Transparent varieties transmit light rays of short wavelengths well; hence their use in spectroscope prisms.

Diagnostic Features: Resembles halite, with which often yields granular intergrowths. Differs from it in taste and by the violet colour which it imparts to the flame. This is visible through a blue filter to filter out the yellow sodium flame. Under the microscope differs from halite in the refractive index.

Readily fusible in the blowpipe flame (at 800°C)′. Addition of $AgNO_3$ to a sylvite solution acidified with nitric acid gives a white AgCl precipitate.

Genesis and Occurrence: Like halite, sylvite is formed in drying salt lakes. However, it is far less common than halite and occurs only in few rock-salt deposits. It is largely adsorbed by soils from migrating surface solutions. Being among the last salts to crystallise from brines, it usually occurs in the upper salt accumulations. Sometimes it is a product of decomposition of carnallite which is formed under the same conditions.

As a sublimation product, it is found on the walls of volcanic craters and in the cracks of solidified lavas. The world's largest deposit of sylvite is at *Solikamsk,* 35 km north of Perm. It was discovered in 1925. The thick lower horizon of sylvinite (halite-sylvite rock) lies at a depth of 150 to 300 m on a body of rock-salt and is overlain by a zone of carnallite mixed with halite. The sylvinite contains from 10 to 35 per cent of KCl. The upper sylvinite horizon is a decomposition product of the carnallite zone (from which $MgCl_2$ has been leached out) and is comprised of a coarse-grained mass of variegated colour in which milk-white sylvite associates with colourless, blue, deep-blue, and greyish rock salt.

Before the Solikamsk discovery, the best known deposits of sylvite were those at *Stassfurt* and other localities in Western and Northern Germany, where salt deposits are of the same Permian age and comprise a rather complex variety of halides, sulphates, and boron compounds. Comparatively younger—Tertiary—sylvinites are mined in *Alsace* (France).

Uses: Potassium salts are predominantly used as fertilizers. Only about 5 per cent are used in the chemical industry for the manufacture of KOH, K_2CO_3, KNO_3, $KClO_3$, $KMnO_4$, KCN, KBr, KI, and other compounds, which have numerous applications: in medicine, photography, in the manufacture of perfumes, fireworks, paper, glass (cut-glass ware and Bohemian glass), varnishes and paints, etc.

Cabnallite, $MgCl_2.KCl\text{-}6H_2O$.

Chemical Composition: Mg 8.7 per cent, K 14.1 per cent, Cl 38.3 per cent, H_2O 38.9 per cent; negligible amounts of isomorphous impurities—Br (up to 0.2 per cent), Rb and Cs (hundredths of per cent), rarely Li and Tl. Common mechanical impurities are NaCl, KCl, $CaSO_4$, Fe_2O_3, clayey matter, droplets of brines, and often abundant occluded gas bubbles (mixture of nitrogen, hydrogen, and methane).

System, orthorhombic; *symmetry,* rhombic dipyramidal. Crystals extremely rare, of pseudohexagonal habit. Commonly granular massive.

Colour, Colourless when pure. Commonly pink crystal, or red owing to inclusions of finely-dispersed iron oxides. Dark brown or yellow when iron hydroxides are present. Lustre vitreous on fresh fracture, on exposure to air quickly becomes greasy.

n_x = 1.494, n_y = 1.475 and n_z = 1.466.

Hardness, 2 to 3. Brittle. *Cleavage,* none. *Specific gravity,* 1.60. Other properties. Unusually hygroscopic. Deliquesces, decomposing to KCl and $MgCl_2 6H_2O$ and forming a thick brine. Taste salty disagreeably bitter. Intensely fluorescent. Dissolving in water is accompanied by a crackle similar to the crunch of footsteps on the frozen snow. This is due to the presence of occluded gas bubbles under high pressure, more frequent in carnallite than in other salts.

Diagnostic Features: Occurs in paragenetic association with rock salt and sylvite. Deliquescent in moist atmosphere. Can be distinguished from bishofite ($MgCl_2 6H_2O$) and tachydrite ($2MgCl_2$;$CaCl_2$;$12H_2O$), which are also deliquescent, only with the aid of microchemical .reactions (contains K), and by the violet colour it imparts to the flame. This is best observed through a blue filter. When scratched with the point of a knife, emits a characteristic crackle, due to sudden expansion of the occluded gas bubbles.

Readily fusible before the blowpipe flame. Like many other water-abundant crystallohydrates, grains of carnallite, when heated on glass, dissolve in their own crystallisation water. On slow drying, small KCl cubes are formed.

Genesis and Occurrence: One of the last salts to be formed from salt-lake brines, it is enriched with potassium and magnesium. Therefore, it occurs in the upper horizons of buried sedimentary salt beds and overlays sylvinite strata. Sometimes segregates on the walls of dry mine workings from the brines which trickle down from overlying strata.

In the USSR occurs in enormous amounts at *Solikamsk* which has already been mentioned. Large masses also found at *Stass-furt* (Germany), *Kalush* (Western Ukraine), and elsewhere.

Uses: Like sylvite and other potassium salts, is used for fertilizers. Electrolysis of dehydrated carnallite in closed baths produces floating metallic magnesium; potassium-enriched residues are utilised in the production of fertilizers.

Refined metallic magnesium goes to make light and strong alloys with aluminium (duraluminium, magnalium) for the aircraft industry, and as a reducing agent for obtaining metallic titanium. Other uses are in flares and flashlight powders for photography, etc.

Bromine extracted from the brines by chemical processes is used in medicine (potassium bromide), photography, etc.

CEBARGYBITE GROUP

This group comprises normal chlorides, bromides, and iodides of Ag, Cu, Hg, and Pb occurring in the oxidised zones of ore deposits under the conditions of hot and arid climate. We shall review only cerargyrite.

Cerargyrite, AgCl. "Ceras" is the Greek for horn, "argyros", silver. Synonym: horn silver (its fracture often appears horny).

Chemical Composition: Ag 75.3 per cent, Cl 24.7 per cent. May contain Hg and Br as impurities.

System, cubic; *symmetry,* hexoctahedral. Crystal structure resembles that of NaCl. Crystals are rare, habit cubic. Twinning along (111). Usually occurs as incrustations, crystalline coatings, horny and waxy sinter masses.

Colour, Freshly-exposed specimens colourless or slightly yellowish, bluish-greenish, and brownish. On exposure to light turns first violet-grey and later even black (possibly owing to segregation of finely-dispersed Ag). *Lustre,* adamantine for crystals and waxy for cryptocrystalline compact masses, $n = 2.07$. *Hardness,* 1.5 to 2. Can be cut with a knife. Malleable. *Cleavage,* none. *Specific gravity,* 5.55. Insoluble in water.

Diagnostic Features: Characterised by low hardness, high refractive index, insolubility in water, and chemical reactions. Before the blowpipe on charcoal fuses readily (450 to 500°) and boils. Roasted with soda in the reducing flame, readily yields metallic silver. Almost insoluble in acids (partly dissolves in HCl). Forms a solution only with ammonia, which is a highly characteristic feature.

Genesis: Occurs in the oxidised zones of silver-lead ore deposits, being formed in the course of oxidation of silver minerals by chlorine-bearing water seeping down from the surface. In larger amounts occurs in localities with a hot and arid climate. Pseudomorphs after native silver have been recorded.

In the USSR, cerargyrite occurs in the oxidised zones of a number of deposits in the Southern Urals (Mikhaylovsky mine in the Baimak district), Kazakhstan, and Altai. The world's largest deposits of cerargyrite and other halide silver compounds are in the Atacama desert (Chile), and also in Bolivia, Mexico and Australia.

Bibliography

Anthony, John W., Bideaux Richard A., Bladh Kenneth W. and Nichols Monte C.: *Handbook of Mineralogy: Elements, Sulfides, Sulfosalts,* Mineral Data Publishing, 1990.

———— : *Mineralogy: Silica, Silicates,* Mineral Data Publishing, 1995.

———— : *Mineralogy: Halides, Hydroxides, Oxides,* Mineral Data Publishing, 1997.

———— : *Handbook of Mineralogy, Volume IV, Arsenates, Phosphates, Vanadates,* Mineral Data Publishing, 2000.

———— : *Mineralogy of Arizona,* University of Arizona Press, Tucson, 1977.

Artini E.: *Mineral,* Hoepli, 1934.

Aubert, G., Guillemin C. and Pierrot R.: *Précis de Minéralogie,* Masson, 1978.

Bancroft, Peter: *The World's Finest Minerals and Crystals,* Viking Press, New York, 1973.

Bates, Robert Latimer: *Our Modern Stone Age,* W. Kaufmann, Los Altos, 1982.

Bayliss, Peter, Erd Dick C., Mrose Mary E., Sabina Ann P. and Smith Deane K.: *Mineral Powder Diffraction File, Data Book, International Centre for Diffraction Data,* 1986.

Betekhtin, A. G.: *Kurs Mineralogii,* Mir, Moskva, 1968.

Beudant F. S.: *Traité élémentaire de Minéralogie,* Verdière, 1830.

Blackburn, William H. and Dennen William H.: *Encyclopedia of Mineral Names*, Mineralogical Association of Canada, 1997.

Bowen, Oliver E.: *Rocks and Minerals of the San Francisco Bay Region*, University of California Press, Berkeley, 1962.

Chesterman, Charles Wesley: *The Audubon Society Field Guide to North American Rocks and Minerals*, Knopf, New York, 1978.

Clark A. M.: *Hey's Mineral Index*, Natural History Museum Publications, Chapman, 1993.

Dana Edward Salisbury, Ford William E., *A Textbook of Mineralogy*, John Wiley, 1949.

De Michele, Vincenzo: *Minerals, their Beauty and Structure*, Orbis Pub., London, 1976.

Deer, W. A.: *Rock-Forming Minerals*, Wiley, New York, 1962.

Deer, W.A., Howie R.A. and Zussman J.: *An Introduction to the Rock forming Minerals*, Longman, London, 1993.

Desautels, Paul E.: *The Mineral Kingdom*, Madison Square Press, New York, 1968.

———————— : *Treasures in the Smithsonian: The Gem Collection*, Smithsonian Institution Press, Washington, 1979.

Eckel, Edwin Butt: *Minerals of Colorado*, Fulcrum Pub., Golden, 1997.

Embrey, Peter G. and Fuller John P.: *A Manual of New Mineral Names 1892-1978*, British Museum, 1983.

Estavillo, William: *Gems and Minerals of California: A Guide to Localities*, Renaissance House, Frederick, 1992.

Fleischer, Michael and Mandarino Joseph A.: *Glossary of Mineral Species 1995*, The Mineralogical Record, 1995.

Gaines, R.W., Skinner H.C.W., Foord E.E., Mason B. and Rosenzweig A.: *Dana's New Mineralogy*, John Wiley, New York, 1995.

Gramaccioli, C.M.: *Minerali alpini e prealpini,* Atlas, 1975.

Grim Ralph E.: *Clay Mineralogy,* McGraw-Hill, London, 1968.

Hausel, W. Dan and Wayne M. Sutherland: *Gemstones and Other Unique Minerals and Rocks of Wyoming: A Field Guide for Collectors,* Wyoming State Geological Survey, Laramie, 2000.

Haüy, René Just: *Traité de Minéralogie,* Bachelier, 1822

Henry, Darold John: *California Gem Trails: A Field Guide for the Collector of California's Gems and Minerals,* L.R. Gordon, Long Beach, 1957.

Hey, Max H.: *A Second Appendix to the Second Edition of An Index of Mineral Species and Varieties Arranged Chemically,* Trustees of the British Museum, 1974.

―――――― : *An Index of Mineral Species Arranged Chemically,* British Museum, 1975.

―――――― : *An Index of Mineral Species and Varieties arranged chemically,* Trustees of the British Museum, 1963.

Kipfer Alex: *Mineral Index,* Ott Verlag, 1974.

Lacroix A.: *Minéralogie de la France et de ses anciens Territoires d'Outre-mer,* Blanchard, 1964.

Ladoo, Raymond B.: *Nonmetallic Minerals,* McGraw-Hill, New York, 1951.

MacFall, Russell P.: *Minerals and Gems: A Color Treasury for Collectors and Guide to Hunting Locations,* Crowell, New York, 1975.

MacLachlan, Don: *Complete Book of Rocks, Gems and Minerals,* Petersen Pub. Co., Los Angeles, 1968.

Mandarino Joseph A.: *New Minerals 1990-1994,* The Mineralogical Record Inc., 1997.

―――――― : *Fleischer's Glossary of Mineral Species 1999,* The Mineralogical Record, 1999.

Merrill, George P. and Foshag William F.: *Minerals from Earth and Sky: The Story of Meteorites, Gems and Gem*, Smithsonian Institution Series, New York, 1934.

Milovsky, A. V. and Kononov O. V.: *Mineralogy*, Mir, 1985.

National Museum of Natural History: *The National Gem Collection*, Smithsonian Institution, Washington, 1997.

Nickel, Ernest H. and Nichols Monte C.: *Mineral Reference Manual*, Van Nostrand Reinhold, 1991.

Oliver E.: *The Care and Conservation of Geological Material: Minerals, Rocks, Meteorites and Lunar Finds*, Oxford University Press, Boston, 1992.

Palache, Berman and Frondel: *Dana's System of Mineralogy*, Wiley & Sons, 1962.

Pearl, Richard Maxwell: *Cleaning and Preserving Minerals*, Earth Science Pub. Co., Colorado Springs, 1982.

Pellant, Chris: *Rocks and Minerals*, Dorling Kindersley, New York, 1992.

Phillips, William R. and Griffen Dana T.: *Optical Mineralogy: The Nonopaque Minerals*, Freeman, 1981.

Picot P. and Johan Z.: *Atlas of Ore Minerals*, BRGM, 1982.

Ramdohr, Paul and Strunz Hugo: *Klockmanns Lehrbuch der Mineralogie*, Enke, 1978.

————— : *The Ore Minerals and their Intergrowths*, Pergamon, 1969.

Ransom, Jay Ellis: *Gems and Minerals of America: A Guide to Rock Collecting*, Harper & Row, New York, 1974.

Reynolds, Robert E.: *Minerals from the Mojave*, San Bernardino County Museum Association, Redlands, 2000.

Robbins, Manuel: *Fluorescence: Gems and Minerals under Ultraviolet Light*, Geoscience Press, Phoenix, 1994.

————— : *Rocks and Minerals*, DK Pub., New York, 1998.

Roberts, Willard L., Campbell Thomas J., Rapp George R. and Wilson Wendell E.: *Encyclopedia of Minerals*, Van Nostrand Reinhold, 1990.

Roberts, Willard L., Rapp George R. and Weber Julius: *Encyclopedia of Minerals,* Van Nostrand Reinhold, 1974.

Rössler Hans J.: *Lehrbuch der Mineralogie,* VEB, 1987.

Sinkankas, John: *Gemstone and Mineral Data Book,* Geoscience Press, Prescott, 1981.

Sofianides, Anna S.: *Gems and Crystals from the American Museum of Natural History,* Simon and Schuster, New York, 1990.

Stalder, H.A., Wagner A., Graeser St. and Stucker P.: *Mineralienlexikon der Schweiz,* Wepf, 1998

Strunz, Hugo and Nickel Ernest H.: *Strunz Mineralogical Tables,* Schweizerbart'sche, 2001.

Strunz, Hugo: *Mineralogische Tabellen,* Geest & Portig, 1982.

Verbeeck: *Biominerals,* CRC Press, Boca Raton, 1990.

Vozza, Vincent J. and Judith L. Vozza: *Zeolites!: The Misunderstood Minerals,* F. Silva Printing, New Jersey, 1975.

Weiss, Stefan: *Das Grosse Lapis Mineralienverzeichnis,* Weise Verlag, 1994.

William, E.: *Familiar Rocks and Minerals,* Knopf, New York, 1988.

Zim, Herbert Spencer: *Rocks and Minerals: A Guide to Familiar Minerals, Gems, Ores and Rocks,* Golden Press, New York, 1957.

Zoltai, Tibor and Stout James A., *Mineralogy, Concepts and Principles,* MacMillan Publishing Company, 1984.

Index

M

N

O

P

Q

S

T

U

W

X

Z

❑❑❑